How to Write & Do Proofs

A Self-Teaching Guide

Clear & Easy to Understand

Dr. Noah

Text Copyright

No part of this publication may be reproduced, or stored in a retrieval system, or transmitted in any form or by any means, electronic, mechanical, photocopying, recording, or otherwise, without written permission of the author.

To my Parents

Preface

I often write many rough notes for myself that I want to think more about, but I have ended up with a lot of notes and no time for thought! Some of my notes are questions and answers about mathematics and proofs. Therefore, this text contains many examples, answers, and discussions about proofs. I tried to give answers and complete discussions of these proof problems.

A proof is just building on other things you know are true. The notes from which this book derives were used in a graduate-level course. The solutions to many of the examples were written over a year. The solutions are provided as a guide only. In some cases, there may be different ways of solving a problem.

Contents

Introduction

One of the most difficult topics in mathematics is writing proofs. If proofs scare you, please allow me to convince you not to be afraid. True wisdom and knowledge comes to each of us only when we realize how little we understand about life, ourselves, and the world around us. Socrates once said, "I know I am intelligent, because I know that I know nothing."

Proofs are a lot like jigsaw puzzles. There are no rules about how jigsaw puzzles must be solved. The only rule concerns the final product. All the pieces must fit together, and the picture must look right. The same is true of proofs.

A proof is just an orderly way to show that something is true, by building on other things you know are true. The only way that order matters is that each thing you say must be based on something you have already said. Often it will be based on the previous statement, but sometimes you will have to use earlier statements as well.

Think of it as building a tower to reach a high goal. Your "givens" are the foundation someone laid for you, and the theorems you have are the girders and rivets you have to put together to make the tower.

- A *proof* is then to give a legitimate (logical) argument or justification why a claim is true, based on the known facts (*Theorems, Propositions, Lemmas, etc*).

- A proof is also a method of communicating a mathematical truth to another person who also "speaks" the language. Therefore, use *mathematical* language as much as you can.

The most two important things a proof must have are the following:

- **Clarity**
- **Backup**

Be sure that your statements are *clear*, and they are each *backed up* with a valid (logical) reason.

To make sure your assertions are adequately justified, you must be skeptical about every inference in your proof. If there is any doubt in your mind about whether the justification you have given for an assertion is adequate, then it is not. After all, if your own reasoning does not even convince you, how can you expect it to convince anybody else?

There is absolutely no reason to write a proof unless your reader can understand what you are saying. Keep this in mind as you write. We always try to pretend we are speaking to someone who has very limited mathematical knowledge and who does not know what facts we have been "given".

Therefore, when we state a property or a theorem or a definition as backup, we always explain it in the proof. We never assume for example, that our reader knows what the second part of the distance postulate is by memory. We also make sure that we state the necessary parts that have been "given" to us. ASSUME YOUR READER KNOWS VERY LITTLE.

Also keep in mind that THERE IS NEVER JUST ONE WAY OF STATING A PROOF.

Do not worry about trying to create a "perfect wording". Just be sure that your statements are clear, and that they are each backed up with a legitimate reason.

A good measure of the quality of your proof is found by reading it to a person who has not taken a geometry course or who has not been in one for a long time. If they can understand your proof by just reading it, and they don't need any verbal explanation from you, then you have a good proof.

A. Know What Exactly You Need to Prove

In order to do a proof, you first must know what exactly you need to prove.

A statement you need to prove is often stated as follows:

"If **A** is true, _then_ **B** is true", or "Assume **A**, then **B**"

In this statement, **A** is called the _Hypothesis_, and **B** is called the _Conclusion_.

For example, in the following statement:

"Let $n = p_1^{a_1} p_2^{a_2} \cdots p_k^{a_k}$, where p_1, ..., p_k, are distinct primes"

Prove that each of a_1, ..., a_2 is even, then **n** is a perfect square.

The first statement:

"$n = p_1^{a_1} p_2^{a_2} \cdots p_k^{a_k}$, where p_1, ..., p_k are distinct primes"

Is the _Hypothesis_ (assumption, or known fact),

And is the second statement of

"**n** is a perfect square"

Is the _Conclusion_, i.e. the part you need to prove.

B. Several Fundamental Proof Techniques

• Proof Technique 1: the <u>Forward-Backward Method</u>

When proving "*A implies B*", you are given that A is true and you must somehow use this information to reach the conclusion that B is true.

When you write the proof, you use the ***forward process***, that is, start from the assumption A to reach the conclusion B by going through several clear and logical arguments. However, in attempting to figure out just how to reach the conclusion, the <u>***trick***</u> is to go through a ***backward process***.

In the ***backward process*** you begin by asking the following *Key Question* (KQ):

   "How or when can I conclude that the statement B is true?"

By asking such question, you can call on the information provided in the assumption, together with your general knowledge (especially the theorems you have learned), thus allowing you to *focus* on those aspects of the problem that *really seem to matter*.

Once you have asked yourself the key question, the next step in the ***backward process*** is to answer it. Sometimes it might take several (backwards) steps to eventually match the assumptions in **A**.

The steps can be as follows:

In order to prove B, we can instead try to prove B_1 the following:

   If B_1 is true, then B is true

Then to prove B_1, we need to prove B_2. We keep moving towards to match the information contained in A.

The ***backward process*** is only some analysis of how to reach the conclusion B. It can be *written on a scratch paper*. However, to write the proof, you need to **translate** your analysis to a rigorous proof.

For example, to write a proof, start from the statement **A**, which is given to be true, and then derive from it by going through several clear and logical steps to conclude that **B** is true. This process is called the *forward process*.

Example 1:

*Show that if **n** is an even integer, then **n²** is an even integer.*

Backward Process: You should ask the KQ:

*"How can I show that **n²** is even?"*

By the definition that we learned as kids, you need to show that **n² = 2k** for some integer **k**.

The problem is then how to find such **k**. Therefore, you need to go back to see what **n** looks like.

We find that we are given that **n** is an even number, that is, **n = 2k**. For example, if k=3, then n=2*3=6, and 6 is an even number. If k=4, then n=8, which is also an even number.

Now if we square **n** and square **2k**, then we get the following:

$$n^2 = 4k_1^2 = 2(2k_1^2)$$

It is still an *even number*.

Is it a proof? **NO**, it is not. But it is an analysis that tells you why **n²** *is an even number* by using the facts that are given (and using the theorems you have learned). You still need to write a clear and logical proof (**forward process**).

Proof of example 1 (Forward Process):

Since **n** is even, we have **n = 2k₁** where **k₁** some integer, then

$$n^2 = 4k_1^2 = 2(2k_1^2)$$

Where **2k²** is still an integer. Hence **n²** is an even integer, and we are done with the proof.

Example 2:

Let $n = p_1^{a_1} p_2^{a_2} \cdots p_k^{a_k}$, where $p_1, \ldots, p_k$ are distinct primes.

Prove that each of $a_1, \ldots, a_k$ is even, then n is a perfect square.

What do you need to prove?

You are asked to prove the following claim or fact:

$$n \text{ is a perfect square}$$

Note the phrase "*perfect square*". For example, 25 is a perfect square of 5, ($5^2=25$), and 27 is a perfect cube of 3, $3^3=27$.

Here n is a perfect square means that $n=m^2$ for some integer m. Hence you **need to show** that $n=m^2$ *for some integer m* (this is statement B).

What is given? Here you are given two facts:

(i) $n = p_1^{a_1} p_2^{a_2} \cdots p_k^{a_k}$, *where* $p_1, \ldots, p_k$ *are distinct primes*

(ii) $a_1, \ldots, a_k$ *are even numbers.* (these are statements A).

What you are supposed to do?

You need to write several steps, based on the facts (i) and (ii), and together with the theorems and results you have learned (use them as you need), in a legitimate way, **show** that *you can find an integer m such that $n=m^2$* (this means that n is a perfect number).

Backward Process: Remember the **goal** is to show that "n *is a perfect number*. This is the same to "*find m such that $n=m^2$*".

How do we find such an **m**? (*KQ: Key Question*)

You need to go back to see what **n** looks like.

We have the following facts:

Fact (i) tells us that $n = p_1^{a_1} p_2^{a_2} \cdots p_k^{a_k}$

Fact (ii) tells us $a_1, \ldots, a_k$ are even numbers

Now do you have any clue how to find m? If not, let us see what does this statement means:

$$a_1, \ldots, a_k \text{ are even numbers}$$

It means that $a_1, \ldots, a_k$ all have number **2** as a factor, or in other words:

$$a_1 = 2b_1, \ldots, a_k = 2b_k \text{ for some integers } b_1, \ldots, b_k$$

We do not know what they are, but we do not care about it, as long as we make sure they are integers.

Now try to plug $a_1 = 2b_1, \ldots, a_k = 2b_k$ into $n = p_1^{a_1} p_2^{a_2} \cdots p_k^{a_k}$.

What we find is the following:

$$n = p_1^{2b_1} p_2^{2b_2} \cdots p_k^{2b_k}$$

And it is the same as:

$$n = (p_1^{b_1} p_2^{b_2} \cdots p_k^{b_k})^2$$

Now do you see what **m** is supposed to be?

Yes! $\quad m = p_1^{b_1} p_2^{b_2} \cdots p_k^{b_k}$

This is how we find **m**.

Is it a proof? It is **NOT**!

But it is an analysis that tells you why **n** *is a perfect number* by using the facts that are given and using the theorems you have learned. You still need to write a **clear** and **logical** proof (**forward process**).

16

Remark: The _bridge_ connecting A to B is the following:

$$n = p_1^{a_1} p_2^{a_2} \cdots p_k^{a_k} \text{,}$$

It links the facts $a_1, ..., a_k$ are even numbers to your destination (i.e. n is a perfect number).

Proof of Example 2 (Forward Process):

Since $a_1, ..., a_k$ are even numbers, $a_1 = 2b_1, ..., a_k = 2b_k$ for some integers $b_1, ..., b_k$.

We also know that $n = p_1^{a_1} p_2^{a_2} \cdots p_k^{a_k}$

Therefore, we are able to do the following:

$$n = p_1^{a_1} p_2^{a_2} \cdots p_k^{a_k} = p_1^{2b_1} p_2^{2b_2} \cdots p_k^{2b_k} = (p_1^{b_1} p_2^{b_2} \cdots p_k^{b_k})^2$$

Then if we let $m = p_1^{b_1} p_2^{b_2} \cdots p_k^{b_k}$, then $n = m^2$.

Hence, n is a perfect number. So the claim is true. And we are done with the proof.

- **Proof Technique 2: the <u>Construction Method</u>**

If you see the words **there is** in **B** you need, in general, to use this technique in the **backward process**.

The idea is to **construct** the desired object (Remember the *construction of the object does not, by itself, constitute the proof*; rather it is used in the **backward process**, the proof needs to show that the object you constructed is in fact the correct one).

How you actually construct the desired object is not at all clear. Sometimes it will be by trial and error; sometimes an algorithm can be designed to produce the desired object-it all depends on the particular problem. In almost all cases, the information in statement **A** will be used to help accomplish the task.

In the proof of example 2 above, we actually used the **Construction Technique**. In fact, to show that:

$$\textit{"n is a perfect number "}$$

It is the same as showing that "there is *an* **m** *such that* **n = m²**. "

We constructed $m = p_1^{b_1} p_2^{b_2} \cdots p_k^{b_k}$ as the proof of Example 2.

Example 3:

Prove that if **a, b** *and* **c** *are integers for which* **a/b** *and* **b/c**, *then* **a/c**.

Backward Process:

What we need to prove is **a/c**. The **KQ** (Key Question) is:

"How can I show that an integer (namely, **a**) divides another integer (namely, **c**)?"

By definition, it is equivalent to show that:

"**there is** an integer **k** such that **c=ak**"

How do we find such **k**?

Use the **Construction Technique**! To find this k, we need to see what a, b and c look like. Well, we are given that a/b and b/c, i.e. $b = ak_1$ $c = bk_2$, where k_1, k_2 are two integers.

Hence, it follows that $c = bk_2 = ak_1k_2$. And so we **found** k!

In fact $k = k_1 k_2$.

Proof of Example 3 (Backward Process):

Since a/b and b/c, we have $b = ak_1$ *and* $c = bk_2$, where k_1, k_2 are two integers.

Hence, it follows that $c = bk_2 = ak_1k_2$, and a/c.

• Proof Technique 3: the <u>Choose Method</u>

If you see the word *for all* in **B**, you need, in general, to use this technique (in the **backward process**). To show that a statement is true *for all* the objects that have the certain property. One way you might be able to do this is to list all of them. When the list is finite, this might be a reasonable way to proceed; however, more often than not, this approach will not be practicable because the list is too long, or even infinite.

The **Choose method** can be thought as a proof machine that rather than **actually** checking that something happens for each and every object with the certain property, has the *capability* of doing so.

The Choose Technique shows you how to design the inner working of the proof machine so that it will have this capability. Here is how it works: You *choose* one object that has the certain property. Then, by using the forward-backward method, you must conclude that, for the chosen object, the something happens. If you are successful, your proof machine will have the capability of repeating the same proof for any of the objects having the certain property.

Example 4:

Let S = { real numbers x: $(x^2 - 3x + 2)$ }, and

T = { real numbers x: $1 \le x \le 2$ }

Prove that $S = T$

Backward Process: The (key) question is:

"How can I show that two sets (namely, **S** and **T**) are equal?"

We need to prove two things:

S is subset of T

and

T is a subset of S

21

To prove that, **S** *is subset of* **T**, the *Key Question* is:

"How can I show that a set (namely, **S**) *is a subset of another?"*

The answer is that you must show the following:

*"For **all** elements* **x** *in* **S**, **x** *is in* **T**"

Thus, it indicates you should proceed with the **choose method**. To do so you must choose an object having the certain property (in our case the certain property is "**x** *in* **S**").

In this case, you **should choose** an element, say **x** *in* **S**. Once you choose such **x**, you need to show such **x** is also in **T** (*this part of analysis is omitted*)!

Proof of Example 4 (forward Process):

We need to prove two things:

S is subset of T

and

T is a subset of S

To see that **S** is subset of **T**, we use the **Choose Technique**.

Let **x** be in **S** (the use of word "let" in condensed proofs frequently indicates that the **choose method** has been invoked!). Consequently, $(x^2 - 3x + 2) \leq 0$, and so $(x - 2)(x - 1) \leq 0$.

This means that either $(x - 2) \geq 0$ and $(x - 1) \leq 0$, or else that, $(x - 2) \leq 0$ and $(x - 1) \geq 0$. The former cannot happen because, if it did, $x \geq 2$ and $x \leq 1$. Hence it must be that $x \geq 2$ and $x \geq 1$, which means that $x \, \varepsilon \, T$.

This proves that **S** is a subset of **T**. The proof that **T** is a subset of **S** is omitted.

• Proof Technique 4: the <u>Specialization Method</u>

The specialization method is used for working forward from a statement that contains *"for all."*

The first step is to identify the element x (or object x), with the certain property, and then the something that happens (for example, $x \varepsilon S$).

The second step is to look for one particular object in which to specialize.

The third step is to verify that the particular object has the certain property, for example, being in the set S.

The last step is to write a new statement in the forward process about the something that happens for that particular object (for example, $y \varepsilon S$).

Steps for Using Specialization

If the statement *"for all"* appears in the <u>*forward process*</u>, use specialization by following these steps:

1.   Identify, the objects, the certain property, and the something that happens.

2.   When the Choose Method is used, search in the <u>*backward process*</u>, for one particular object to apply specialization to.

3. Verify that this object in step 3 **does** have the certain property specified in the 'for all' statement.

4. Finish by writing a new statement in the forward process, that the 'something' happens for this one object.

Be Careful not to confuse the *Choose Method* with the *Specialization Method*. Use the Choose Method when you find the key words *"for all"* in the <u>*backward process*</u>; use the Specialization Method when you see the key words *"for all"* in the forward process.

• Proof Technique 5: the <u>Induction Method</u>

A separate proof of technique, which is known as **mathematical induction**, is to prove the statement **B** containing "*for all*".

Induction can be confusing at first, but once you get the hang of it, it is easier. You will probably remember how to do proofs by induction if you have a good understanding of how the proofs work.

Proofs by induction are commonly used when you want to prove a statement that depends on some variable (usually named n) for all positive integer values of that variable.

For instance, you want to prove

$$|A^n| = |A|^n$$

The above equality for all positive integer values of n. In other words, you want to show that the above equality holds for n = 1, n = 2, n = 3, and so on.

How does the inductive proof do this? First, it demonstrates that the statement holds for the "base case" - usually the n = 1 case.

Next, when doing an inductive proof, you assume that the statement holds for the kth case. What is the kth case, you may be wondering? It is any case. We write k because we want k to be able to represent any positive integer.

Next, you prove that the (k+1)st case holds assuming that the kth case holds. Then the proof is done! Why? Well, think about it. In the first part we showed that the statement we are trying to prove holds for n = 1, right? Then, in the second part of the proof, we demonstrate that if the kth case holds, so does the (k +1)st case. Combining these two pieces of information, we see that since the 1st case holds (as proven in the first part of the proof), the 2nd case must hold too (according to the second part of the proof); well, if the 2nd case holds (as just demonstrated), so too must the third case hold, right? And so on. Thus we have shown that the statement holds for any *n*.

This will all become much clearer if we do an example. Let's look at this example. To show that $|A|^n = |A^n|$ for all n by induction, we follow the steps outlined above.

1. Show that the statement holds for the base case **n = 1**.

When $n = 1$, the equation is $|A|^1 = |A^1|$, right?

Does this equation hold? Yes, because anything to the first power is itself, so $|A|^1 = |A|$ and $|A^1| = |A|$. Hence, $|A|^1 = |A^1|$. Thus, we have shown that the equality holds for $n = 1$.

2. Now we assume that the equality holds for the kth case. That is, we assume that $|A|^k = |A^k|$.

3. We now want to prove that the (k+1)st case holds.

We want to prove that $|A|^{k+1} = |A^{k+1}|$ (**). Let us use what we know. In inductive proofs, proving that the $(k+1)^{st}$ case holds almost always relies on the fact that we have assumed that the kth case holds. So let's rewrite the equation for the $(k+1)^{st}$ case in a way that will allow us to use information from the kth case.

We know that $|A|^{k+1} = |A|^k * |A|$, right? Similarly, we have that

$$|A^{k+1}| = |A^k * A| = |A^k| * |A|.$$

Substituting into (**), we see that we want to prove that:

$$|A|^k * |A| = |A^k| * |A|$$

Once you have proved that the above equation holds, you will have proven that the (k+1)st case holds, and thus, the proof will be over.

Again, the reason the proof works is as follows: First, we prove that the equation holds for $n = 1$. Next, we prove that if it holds for the kth case then it must also hold for the $(k+1)^{st}$ case. Since the equation holds for $n = 1$, it must also hold for $n = 2$ (the second part of the proof shows that if the kth case holds then so must the $(k+1)^{st}$. Here, $k = 1$). But if the equation holds for $n = 2$, it must also hold for $n = 3$, and so on.

• Proof Technique 6: the <u>Contradiction Method</u>

With all the proof techniques you have learned so far, you may well find yourself unable to complete a proof for one reason or another. As powerful as the **forward-backward method** is, it may not always lead to a successful proof.

When trying to prove a statement is true, it may be beneficial to ask yourself, "What if this statement was not true?" and examine what happens. This is the premise of the Indirect Proof or Proof by Contradiction.

<u>Steps in an Indirect Proof</u>: *(Proof by Contradiction)*

1. Assume that the opposite of what you are trying to prove is true.

2. From this assumption, see what conclusions can be drawn. These conclusions must be based upon the assumption and the use of valid statements.

3. Search for a conclusion that you know is false because it contradicts given or known information. Oftentimes you will be contradicting a piece of GIVEN information.

4. Since your assumption leads to a false conclusion, the assumption must be false.

5. If the assumption (which is the opposite of what you are trying to prove) is false, then you will know that what you are trying to prove must be true.

How to recognize when an indirect proof is needed: generally, the word "*not*" or the presence of a "*not symbol*" (such as the not equal sign ≠) in a problem indicates a need for an **Indirect Proof.**

Example: In the diagram below, $\triangle ABC$ is *not* isosceles.
Prove that if altitude $\overline{BD}$ is drawn, it will *not* bisect $\overline{AC}$.

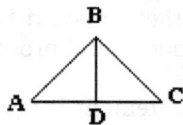

STATEMENTS		REASONS	
1.	$\triangle ABC$ is not isosceles altitude $\overline{BD}$	1.	Given
2.	Assume $\overline{BD}$ bisects $\overline{AC}$	2.	Assumption leading to a contradiction.
3.	D is midpoint of $\overline{AC}$	3.	Bisector of a segment divides the segment at its midpoint.
4.	$\overline{AD} \cong \overline{DC}$	4.	Midpoint divides a segment into two congruent segments.
5.	$\overline{BD} \perp \overline{AC}$	5.	The altitude of a triangle is a line segment extending from any vertex of a triangle perpendicular to the line containing the opposite side.
6.	$\angle ADB, \angle BDC$ right angles	6.	Perpendicular lines meet to form right angles.
7.	$\angle ADB \cong \angle BDC$	7.	All right angles are congruent.
8.	$\overline{BD} \cong \overline{BD}$	8.	Reflexive Property
9.	$\triangle ADB \cong \triangle CDB$	9.	SAS
10.	$\overline{AB} \cong \overline{BC}$	10.	CPCTC
11.	$\triangle ABC$ is isosceles	11.	An isosceles triangle is a triangle with two congruent sides.
12.	$\overline{BD}$ does not bisect $\overline{AC}$	12.	Contradiction steps 1 and 11

Let us look at another example:

*Prove that if **n** is an integer and **n²** is even, then **n** is even.* The **forward-backward method** gives rise to this ***Key Question***:

*"How can I show that an integer (namely, **n**) is even?"*

One answer is to show that "there is an integer **k** such that **n** = **2k**.

The word "there is" suggests the **construction** method. How to find(construct) such **k** ? Look back to see what we are given. We are given that **n²** is even, so $n^2 = 2k_1$ for some positive integer **k₁**. It natural to take the square root to obtain $n = \sqrt{2k_1}$. But how can you rewrite $n = \sqrt{2k_1}$ as $n = 2k$? It would seem that the **forward-backward method** has failed!

Fortunately, there are several other techniques that you might want to try before you five up. One of them is the **Proof by Contradiction Method**. This is how it works: with the **Proof by Contradiction Method**, you begin by assuming that **A** is true, just as you would do in the forward-backward method.

However, to reach the desired conclusion that **B** is true, you proceed by asking yourself a simple question "**Why can't B be false?**" After all, if **B** is supposed to be true, then *there must be some reason why **B** cannot be false.* The **objective** of the method is to **discover** that reason.

In other words, the **idea** of the **Proof by Contradiction Method** is to assume that **A** is true and **B** is **false**, and see why this cannot happen. So what does it mean to "see why this cannot happen"? Suppose, for example, that as a result of assume that **A** is true and **B** is **false**, you were somehow able to reach the conclusion that **1** = **0**! Would that not convince you that it is impossible for **A** is true and **B** is **false** simultaneously?

Thus, *in a Proof by Contradiction, you assume that **A** is true and **B** is **not** true. You must use this information to reach a* **contradiction** *to something that you absolutely know to be true.*

Example 5:

*Prove that if **n** is an integer and **n²** is even, then **n** is even.*

Backward Process: You use the **Proof by Contradiction Method**. So you assume that **A** is true and **B** is NOT true. In this case, you assume that **n** is an integer, **n²** is even (**A**) and **n** is NOT even (**B** is NOT true).

A **contradiction** must now be reached by using this information. Let us analyse how to reach the contradiction. First we use the information that **n** is not even, so **n** is odd. Hence, we can write it as **n = 2k + 1**. Now we want to the information about **n²**.

So we take the square for **n = 2k + 1** to get **n² = 4k² + 4k + 1**. By the information about **n²**, we also know that **n²** is even.

Therefore, **2|n²**. Hence **2|(4k² + 4k + 1)**. It follows that **2|1** which is a **contradiction!**

Proof of Example 5 (forward Process):

We use the method of ***Proof by contradiction***. Assume, to the contrary, that **n** is not even. Then **n** is odd. Hence we can write it as $n = 2k + 1$. Hence $n^2 = 4k^2 + 4k + 1$. By the assumption, we also know that n^2 is even. So $2|n^2$. Hence, $2|(4k^2 + 4k + 1)$. It follows that $2|1$ which is a **contradiction**. This contradiction establishes the claim.

C. The 'if and only if' statements.

Some of the most useful statements in mathematics come in the form of an *"if and only if"* statement. In this case, it asks you to prove **both** directions. For example, in the statement:

*"A holds if and only if **B** holds"*

It asks you to prove

*"if **A**, then **B**" and "if **B**, then **A**"*

D. Several Warnings!

- **Examples supporting a statement do not constitute a proof!**

Here is an example, if you are asked to prove the following:

"The sum of any two even integers is always an even integer"

And you wrote this:

$2 + 4 = 6 \quad 28 + 12 = 40 \quad 123456 + 765432 = 888888$

"Hence the sum of any two even integers is always an even integer."

This is **NOT** a proof! This only helps you to increase your confidence to believe that this statement **should** be true.

- **Use the mathematical languages as much as you can!**

Here is an example, if you are asked to prove the following:

"The sum of any two even integers is always an even integer"

And you wrote like this:

*"Any even integer has **2** as its divisor, so when you add two even integers, we still get a number which has 2 as a divisor, hence it is still an even integer"*

Is the proof correct? *Maybe.* But you better write it by using mathematical languages as much as you can.

Here is what you should write:

Let n_1, n_2 be two even integers.

Then,

$$n_1 = 2k_1, n_2 = 2k_2$$

Therefore,

$$n_1 + n_2 = 2(k_1 + k_2)$$

which is also an even integer.

This proves the statement that the sum of any two even integers is always an even integer.

E. More examples

Example 6:

Suppose that $n = p_1^{a_1} p_2^{a_2} \cdots p_k^{a_k}$, where $p_1, \ldots, p_k$ are distinct primes. Prove that n is a perfect square if and only if each of $a_1, \ldots, a_k$ is even.

This problem actually consists of two problems:

Problem (I): Suppose that $n = p_1^{a_1} p_2^{a_2} \cdots p_k^{a_k}$. Prove that if n is a perfect square, then each of $a_1, \ldots, a_k$ is even.

In this problem, the facts that are given are:

(i) $n = p_1^{a_1} p_2^{a_2} \cdots p_k^{a_k}$

(ii) n is a perfect square

The claim you need to prove is: "each of $a_1, \ldots, a_k$ is even".

Problem (II): Suppose that $n = p_1^{a_1} p_2^{a_2} \cdots p_k^{a_k}$. Prove that if each of $a_1, \ldots, a_k$ is even, then n is a perfect square.

In this problem, the facts that are given are:

(i) $n = p_1^{a_1} p_2^{a_2} \cdots p_k^{a_k}$

(ii) each of $a_1, \ldots, a_k$ is even.

The claim you need to prove is:

"n is a perfect square"

The problem (II) has been proved in Example 1. We now consider problem (I).

33

What you need to prove?

You need to prove that you can find some distinct prime numbers $p_1, \ldots, p_k$ and even numbers $a_1, \ldots, a_k$ such that $n = p_1^{a_1} p_2^{a_2} \cdots p_k^{a_k}$.

What are given?

You know (you are given) that **n** is a perfect number. Also please do not forget to use the theorems that you have learned (all these theorems also become facts!).

What you are supposed to do?

You need to write several steps (statements), based on the fact that **n** is a perfect number and the theorems you have learned, to **show** that $n = p_1^{a_1} p_2^{a_2} \cdots p_k^{a_k}$, where $p_1, \ldots, p_k$ are distinct primes and $a_1, \ldots, a_k$ are even numbers.

Backward Process: Remember that your *goal* is to show that $n = p_1^{a_1} p_2^{a_2} \cdots p_k^{a_k}$, where $p_1, \ldots, p_k$ are distinct primes and $a_1, \ldots, a_k$ are even numbers.

What you know (you are given) is that **n** is a perfect number. Of course, this piece of information seems not enough. What do we do? We need help! Where do we get help? From the results we have learned!

Think it about carefully: the statement you need to prove involves **factorization**! Agree? If you agree, then let us find what result we have learned about **factorization**. It is the following result.**Fundamental Theorem of Arithmetic**: *each integer $n > 1$ can be expressed as a product of powers of primes in the form $n = p_1^{a_1} p_2^{a_2} \cdots p_k^{a_k}$, where $p_1, \ldots, p_k$ are distinct primes and $a_1, \ldots, a_k$ are positive integers."*

Apply the **Fundamental Theorem of Arithmetic** to **n** and we get $n = p_1^{a_1} p_2^{a_2} \cdots p_k^{a_k}$. Are we done? No!

In fact, if you compare it with "what we need to show", you'll see that it remains to be shown that $a_1, \dots, a_k$ are even numbers. How do we show that? We still need to use "what we are given":

n is a perfect number, i.e., $n = m^2$ *for some integer* **m**.

How do we use it?

Well, here is the **trick**: since $n = m^2$, why don't we apply **Fundamental Theorem of Arithmetic** to *m*.

NOTE: we want to get **powers that are even**.

Let us try this: applying the **Fundamental Theorem of** to *m*, we have $m = p_1^{b_1} p_2^{b_2} \cdots p_k^{b_k}$

NOTE: since eventually we need $n = p_1^{a_1} p_2^{a_2} \cdots p_k^{a_k}$, here we use $b_1, \dots, b_k$ as powers for *m*).

Then $n = m^2 = (p_1^{b_1} p_2^{b_2} \cdots p_k^{b_k})^2 = p_1^{2b_1} p_2^{2b_2} \cdots p_k^{2b_k}$.

We got it! Since $a_1 = 2b_1, \dots, a_k = 2b_k$, they are even numbers!

Remark: Here I think the bridge is the **Fundamental Theorem of Arithmetic** which links " n is a perfect number" and the destination " $n = p_1^{a_1} p_2^{a_2} \cdots p_k^{a_k}$, where $p_1, \dots, p_k$ are distinct primes and $a_1, \dots a_k$ are even numbers". The facts we used are:

(i) *n* is a perfect number

(ii) **Fundamental Theorem of Arithmetic**. Based on the analysis, the writing of proof becomes quite simple.

Poof of Example 6 (forward process):

This is an *'if and only if'* statement. This problem actually consists of two problems:

Problem (I) ($\Rightarrow$): *Suppose that* $n = p_1^{a_1} p_2^{a_2} \cdots p_k^{a_k}$. *Prove that if* n *is a perfect square, then each of* $a_1, \ldots, a_k$ *is even.*

Problem (II) ($\Leftarrow$): *Suppose that* $n = p_1^{a_1} p_2^{a_2} \cdots p_k^{a_k}$. *Prove that if each of* $a_1, \ldots, a_k$ *is even, then* **n** *is a perfect square.*

The " $\Leftarrow$ " part has been proved in Example 2. Therefore, we only need to prove the " $\Rightarrow$ " part. In fact, by assumption (the fact we are given), n is a perfect number. So $n = m^2$, where **m** is an integer. We can assume that **m** is a positive integer(since $n = m^2 = (-m)^2$).

If $m = 1$, then $n = 1 = p^{2 \times 0}$ so the claim is true. Now assume that $m > 1$, then by the **Fundamental Theorem of Arithmetic**, $m = p_1^{b_1} p_2^{b_2} \cdots p_k^{b_k}$, where $p_1, \ldots, p_k$ are distinct primes and $b_1, \ldots, b_k$ are positive integers.

Hence, $n = m^2 = (p_1^{b_1} p_2^{b_2} \cdots p_k^{b_k})^2 = p_1^{2b_1} p_2^{2b_2} \cdots p_k^{2b_k}$.

Let $a_1 = 2b_1, \ldots, a_k = 2b_k$, then $a_1, \ldots, a_k$ are even numbers and $n = p_1^{a_1} p_2^{a_2} \cdots p_k^{a_k}$. Therefore, the claim holds. This finishes the claim.

Chapter 1
Proofs and Answers

1.2   Which of the following statements are true?

a.   The cube root of any integer number is a real number

The statement is **TRUE.**

Proof: By **definition** of **cube root**, the cube root of an integer is this integer raised to the exponent $(1/3)$.

For Example, the cube root of $N = N^{1/3}$

And by definition of "exponential", $N^{1/3}$ is **always defined** and it is a **real number** equal to:

$$e^{(\ln(N) * (1/3))}$$

For example, $(8)^{1/3} = e^{(\ln(8) * (1/3))} = e^{.69314718} = 2$

1.2b   For every angle t, $\sec^2(t) - (\tan^2(t)) = 1$

The statement is **FALSE.**

Proof: To prove that something is **false,** we will give a counter-example. Let us consider the angle $t = pi/4$. So we have that $\sec^2(pi/4) = \cos^2(pi/4) / \sin^2(pi/4)$.

And as $\cos(pi/4) = \sin(pi/4) = \mathrm{sqrt}(2) / 2$, then $\cos^2(pi/4) = \sin^2(pi/4) = \frac{1}{2}$.

And so, $\sec^2(pi/4) = (\frac{1}{2}) / (\frac{1}{2}) = 1$

Then using the same reasoning,

$$\boldsymbol{tan^2(pi/4)} = sin^2(pi/4) / cos^2(pi/4) = (1/2) / (1/2) = 1$$

In conclusion $\sec^2(pi/4) - \tan^2(pi/4) = 1 - 1 = 0$

Therefore, for $t = pi/4$, the statement is false. And it is false that $\sec^2(t) - \tan^2(t) = 1$ for all t.

1.2c $x^2 + y^2 > 1$ (where y and x are real numbers)

The statement is **FALSE**.

To prove it, we will try to find a counter-example.
Let us consider the case $x = y = 0$.
Both x and y are real numbers (because 0 is a real number), but
$x^2 + y^2 = 0 < 1$
So we have found a counter-example that when $x = y = 0$, it is false that $x^2 + y^2 > 1$
So the statement is **FALSE.**

1.2d   If $x > 0$, then the $\log_7$ of (x) is greater than zero
(where x is a real number)

The statement is **FALSE**.
To prove it, we will try to find a counter-example.
Let us consider the example $x = \frac{1}{2}$

For this case, x > 0, but $\log(x) = \log(1/2) = -\log(2) < 0$
So for $x = \frac{1}{2}$, x > 0 but $\log(x) < 0$

So we have found a counter-example for the statement above, and so the statement is **FALSE.**

1.5   If I do not get my bike fixed, I will miss my job interview," says Peter. Later, you come to know that Peter's bike was repaired but that he missed his job interview. Was Peter's statement true or false?

On first examination, we do not know if Peter's statement is true or false. It may be both.

First of all, we may think at first that Peter was wrong, because he said that if the bike was not repaired then he will miss his job interview, then he repaired the bike, but he anyway missed the interview. But Peter had not said: "*that if the bike was repaired then he will not miss the interview.*"

He only said that, "*if his bike was not repaired, then he would miss his interview.*" And we do not know if he was right, because as the bike was repaired, it is impossible to confirm if the statement was true or false.

If Peter had not repaired the bike, we would be able to tell whether his statement was true or false. However, the bike *was* fixed, and so we do not know if he was right or not. We would have to be able to consider all the possible combinations (bike fixed, bike non fixed, job interview missed, job interview not missed) to decide. So we can not conclude that the statement was false, and also we can not conclude that it was true.

1.7   Suppose someone says to you that the following statement is true: "If Peter is younger than his father, then Peter will not lose the contest." Using this table did Peter win the contest? Why or why not?

A	B	A implies B
T	T	True
T	F	False
F	T	T
F	F	T

The statement, *"If Peter is younger than his father, then Peter will not lose the contest"* is the same as saying *"A implies B"*. *"If A then B"* is as another way of saying *"A implies B"*.

Here, *"A"* is the first condition, and *"B"* is the consequence, so we can translate the original phrase to a more logically-related form:

"Peter is younger than his father IMPLIES Peter will not lose the contest"

Here, **A** is like *"Peter is younger than his father"* and **B** is *"Peter will not lose the contest"*

Now the table tells us when we can consider *"A implies B"* true, depending on if A and B are true.

We know just one thing from the question. We know that *"If Peter is younger than his father, then Peter will not lose the contest"* is TRUE.

So from the table above, we know that *"A implies B"* will be TRUE. This means that the second row in the table, where *"A implies B"* is false, must be deleted from the table, as it is not compatible with the problem. Therefore, let us eliminate this row, as it is impossible to our problem.

Then we have:

A	B	A implies B
T	T	True
F	T	T
F	F	T

Now we know that *"Peter is younger than his father"* is always TRUE since the statement did not say Peter was adopted. So we know that the statement *"A"* is TRUE (because *"A"* is "Peter is younger than his father"). So again, we look at the table and eliminate all the rows where the value under *"A"* is not true.

And we eliminate the last two rows. We then obtain a table with a single row:

A	B	A implies B
T	T	True

Now we have the acceptable combination of values for *A, B and "A implies B"* that satisfies the conditions in the problem.

So we conclude that *B* is true.

Therefore, **"Peter will not lose the contest"** is true.

So we can conclude that "Peter will not lose the contest" is true.

1.9   By considering what happens when *A* is true and when *A* is false, it was decided that when trying to prove this statement "*A implies B*" is true, you can assume that *A* is true and your goal is to show that *B* is also true. Use the same type of reasoning to derive another method for proving that "*A* implies *B*" is true by considering what happens when *B* is true and what happens when *B* is false.

The problem said if we want to know that "*A implies B*" is true, we can assume that *A* is true and then try to prove that *B* is true.

TABLE:

A	B	A implies B
T	T	True
T	F	False
F	T	T
F	F	T

Now we start working from *B*. What we want to avoid is the combination of the second row happening. For example, we want to prove that the combination *A* = True, *B* = False is not happening. We will consider *B*, and then we will look at *A* to see what is happening.

Therefore, we want to avoid, if *B is False*, then *A is True*. For what we need is that, *if B is False*, then *A is also False*. We proceed by assuming that *B* is false, and we prove that *A* is also false.

Then to prove that "*A implies B*" is True, we assume that *B* is False, and then prove that *A* must be also False.

1.12   Suppose you must show that "*A* implies *B*" is false.

a.   According to the above table, how should you proceed? What should you try to show about the truth of *A* and *B*?

According to the previous table:

A	B	A implies B
T	T	True
T	F	False
F	T	T
F	F	T

"*A implies B*" is only false in the second row when *A* is True and *B* is False. So to prove that "*A implies B*" is False, we should prove that *A* is True, and that *B* is False.

b.   Apply your answer to part (a) to show that the following statement is false: "If $x > 0$, then $\log_7 (x) > 0$"

We want to show that "*If $x > 0$, then $\log_7(x) > 0$*", by applying the rule obtained in (a).

Since "*If A then B*" is the same as "*A implies B*", and to prove that "*A implies B*" is false, we MUST prove that *A* is True, and that *B* is False. We will find an example where

A is True, for example, $x > 0$ is True.
B is False, for example, $\log_7 (x) > 0$ is False.

We need to find a value of x such that $x > 0$ is True and $\log_7 (x) > 0$ is False. We want find a counter-example to prove that the statement is False.

Let us consider $x = 1/7$

Then for $x > 0$, *A* is True.
And $\log_7(x) = \log_7(1/7) = -\log_7(7) = -1 < 0$, *B* is False.

So *A* is True and *B* is False, and "*A implies B*" is False. Therefore, we have proved that the statement is false. Hence, we have found a counter-example; a case when A is True, but B is False.

Chapter 2
The Forward and Backward Method

2.2   When *formulating* the key question, should you look at the last statement in the forward, or the backward process? When *answering* the key question, should you be guided by the last statement in the forward, or the backward process?

When we formulate the *Key Question*, we will look at the first statement in the forward process, and we will also look at the last statement in the backward process. That is, we will look at the conditions. Therefore, we should look at the last statement in the backward process.

When we answer the key question, we will look at the last statement in the forward process, and we will look at the first statement in the backward process (that is, we will look at the goals). So we should be guided by the last statement in the forward process.

2.4   For the key question, "How can we show that two lines in a plane are parallel?" which of the following answers is not correct?

a. Show that the slopes of the two lines are the same
b. Show that the each of the two lines is parallel to a third line
c. Show that each of the two lines is perpendicular to a third line
d. Show that the lines are on opposite sides of a quadrilateral

The answer *d* is incorrect, because showing that the lines are opposite does not mean that they are parallel. Not all quadrilaterals have opposite sides that are parallel. The fact is that quadrilaterals that verify this condition are called "parallelograms", but *not* all *quadrilaterals are parallelograms*.

In the picture above, we can see a quadrilateral with two opposite sides that are not parallel. This picture is a counter-example for the case d. Therefore, this shows that d is not correct.

a is correct by the definition of slope. If the slopes of the lines are the same then the lines are parallel. b is correct because the property of parallelism is a transitive property, meaning that if A is parallel to B and B is parallel to C, then A is parallel to C. So proving that 2 lines are parallel to a third line is sufficient to prove that the two lines are parallel between themselves. c is also correct.

2.6   Suppose we are trying to prove that, "if l_1 and l_2 are parallel lines intersected by a third line l_3, then the smallest angle, S, formed by the intersection of l_1 and l_3 is equal to the smallest angle, T, formed by the intersection of l_2 and l_3." What is wrong with the key question, "how can we show that two parallel lines intersect a third line?

What is wrong is that we are trying to show something that we already know. We are trying to show that "*two parallel lines intersect a third line*", but this is already assumed, since the problem said, "*if l_1 and l_2 are parallel lines intersected by a third line l_3*". So why prove something that is already stated by the problem? We are trying to prove the wrong thing.

What we should be trying to prove is something that will give us the question of the problem, that is, to prove that the smallest angle S formed by l_1 and l_3 are equal to the smallest angle T formed by l_2 and l_3. This has nothing to do with proving that l_1 and l_2 intersect l_3.

2.9   For each of the following, list as many key questions as you can. Be sure your questions contain no notation or symbols from the specific problem, that is, all the symbols are spelled out in English.

a.   If a and b are positive, real numbers, then
$a^2 + b^2 \le [a+b]^2$

Q1: How can I show that the sum of the squares of two non-negative numbers is less-than-or-equal-to the square of its sum?

Q2: What can I prove about the addition of the squares of two non-negative numbers?

Q3: What condition must satisfy two numbers such that the sum of its squares is less-than-or-equal to the square of its sum?

Q4: If I know that two numbers satisfy *"the sum of its squares is less-than-or-equal* to *the square of its sum"*, what else can I say about these numbers?

b.   If $y=m_1x + b_1$, and $y = m_2x + b_2$ are two equations of two lines for which $m_1 = m_2$, then the two lines are parallel.

Q1: What can I say about two lines knowing that they have equations with the same x coefficient?

Q2: What can I conclude about the slopes of two lines knowing that they have equations with the same x coefficient?

Q3: What can I conclude about the parallelism of two lines, knowing that they have the same slope?

Q4: How can I prove that two lines are parallel? What conditions must be satisfied?

c.   If RST is an isosceles triangle with sides ST, RS, and TR, and SU is a perpendicular bisector of RT, then ST=RS

Q1: What can I say about two of the sides of a triangle knowing that the triangle is isosceles?

Q2: If a perpendicular bisector of one side of an isosceles triangle passes by the opposed vertex, what can be said about the two other sides?

Q3: If one line is a perpendicular bisector of a segment, what can be concluded about all the points of this line in respect to the extremes of the segment?

Q4: How can I prove that the two sides of a triangle are equal knowing that the triangle is isosceles, and that the perpendicular bisector of the third side passes by a vertex of the triangle?

Q5: Knowing that a perpendicular bisector of a side in a triangle passes by the third vertex, what can I conclude about the two other sides of the triangle?

d.   If T and R are the sets described below, then R intersects T is a singleton (a set with only one element)

If $R = \{$ real numbers $x : x^2 - x <= 0 \}$,
             $S = \{$ real numbers $x : -(x-1)(x-3) <= 0\}$, and
             $T = \{$ real numbers $x : x >= 1 \}$

Then R intersects S is a subset of T."

The hypothesis here is that $A = $ "R, S, T"
The conclusion here is that $B = $ "*the intersection of* R *and* T *is a singleton*"

Q1: How can I prove that the intersection of R and T is not empty?

Q2: How can I prove that the intersection of R and T has no more than 1 element?

Q3: What elements are at the intersection of R and T?

Q4: For what values of x is it true that $x^2 - x \leq 0$?

Q5: How could I prove that the set $x^2 - x \leq 0$ intersects the set $x \geq 1$ in a single point?

Q6: How can I prove that the intersection of R and S is a subset of T?

Q7: How can I prove that if x is in R and if x is in S, then $x \geq 1$?

Q8: What are the elements of the set "R intersection S"?

Q9: Assuming R intersection S is included in T, & knowing that T is given by $x \geq 1$

Q10: How could I prove that if x is in R intersection S, then $x \geq 1$?

Q11: How could I prove that if $x^2 - x \leq 0$ and $-(x-1)(x-3) \leq 0$, then $x \geq 1$?

<u>2.12</u>   For each of the following key questions, list as many answers as possible

<u>a.</u>   How many can we show that two lines are parallel?

1. We can show that two lines are parallel, by showing that they do not intersect at any point

2. We can show that two lines are parallel, by showing that both are perpendicular to a third line

3. We can show that two lines are parallel, by showing that they have the same slope

4. We can show that two lines are parallel, by showing that they are the opposite sides of a parallelogram

5. We can show that two lines are parallel, by showing that they are both parallel to a third line

<u>b.</u>   How can we show that a set is a subset of another set?

1. We can show that a set is a subset of another set, by showing that all the elements of the first set are elements of the second set

2. We can show that a set is a subset of another set, by showing that the intersection of both sets is the first set

3. We can show that a set is a subset of another set, by showing that the union of both sets is the second set

4. We can show that a set is a subset of another set, by showing that the first set does not intersect with the complement of the second set

5. We can show that a set is a subset of another set, by showing that the first set is included in a third set, that is already included in the second set

2.14   For triangles *SUT and RSU* in the figure below, suppose you have asked the key question; "How can we show that two triangles are congruent?" What is wrong with the answer, "Show that Angle *RUS* = Angle *SUT*,   Angle *RSU* = Angle *UST*, and Angle *SRU* = Angle *STU*?"

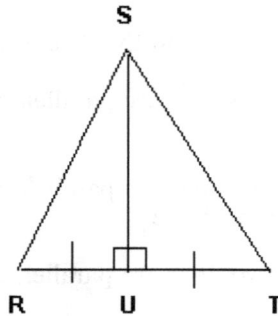

For example, we know from the figure that *RU* = *UT* (segments) and *SU* is a common side for both triangles. So based on the figure, we should use the following answer to the key question above:

Show that RU = UT, < RUT ≤ TUS and US = US

For example, we can show that two triangles are congruent by showing that two sides are equal and that the angles formed are equal.

2.16   For each of the following hypotheses, list as many statements as possible that are a result of applying the forward process one step from the hypothesis.

a.   Rectangle $ABCD$ is a square

1. $AB = CD = AD = BC$
   (By the definition of a square, the 4 sides are equal)

2. AB is perpendicular to BC, AD is perpendicular to CD
   ([By the property of square, the adjacent sides are perpendicular)

3. AB is parallel to CD, AD is parallel to BC
   ([By property of square, opposite sides are parallel)

4. $AC = BD$
   (By property of square, the diagonals are equal)

5. AC perpendicular to BD
   (By property of square, the diagonals are perpendicular)

b.   Integer $n^2 - 1$ is odd

1. The integer n^2 is even
   (Because if $n^2 - 1$ is odd, then n^2 must be even)

2. The integer $n^2 + 1$ is odd
   (Because if $n^2 - 1$ is odd, then $n^2 + 1$ is odd)

3. The integer n is even
   (Because n^2 is even, so n must be even, as the square of an odd number is also an odd number)

4. The integer $(n - 1)$ is odd
   (Because $(n-1)(n+1) = (n^2 - 1)$, and $(n-1)$ and $(n+1)$ being at a distance of 2, they both must be odd or even; but two even numbers multiplied by themselves give other even number, so the unique alternative is that both $(n-1)$ and $(n+1)$ are odd)

c.   Line $y = 3x - 1$ is tangent to the function $x^2 + x$

1. There is a point $(x0, y0)$ in the graph of the function $y = x^2 + x$ such that the tangent line at $(x0, y0)$ has the equation: $y = 3x - 1$

2. There is a point xo such that "the derivative of $x^2 + x$ at $x0$" = 3

3. There is a point xo such that $2x0 + 1 = 3$ and that $3x0 - 1 = x0^2 + x0$

2.18   Suppose we are trying to prove that, "*If R is a subset of S and S is a subset of T, then R is a subset of T,*" What is wrong with the following statement in the forward process? "Because R is a subset of S, it follows that every element of S is in also an element of R."

The statement is FALSE, because if R is a subset of S, then it follows that every element of R is also an element of S, and not the opposite, as the statement ascertains.

Also the statement is wrong for the backward method, because the backward method should start considering the conclusion, that is, we should start considering what must happen to be able to conclude that "*R is a subset of T*".

2.21   Consider the problem of proving that, "If RST is the triangle below, then triangle SUT is congruent to triangle SUR." For this condensed proof, write an analysis indicating the forward and backward steps, and the key questions and answers.

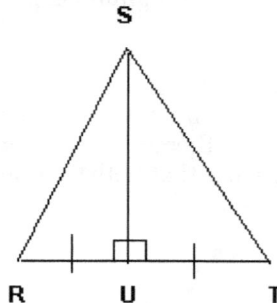

Q1: How can I prove that *SUR* and *SUT* are congruent triangles, knowing that they are as in the figure?

We then see what we know: we know that *SUR* and *SUT* have a common side, that is *SU*, and we also know from the figure that angle $SUR \leq SUT = 90^\circ$, so they have a common angle.

Q2: How can I prove that two triangles are congruent, knowing that they have one angle and one side in common?

If two triangles have one angle in common, and two sides (adjacent to the angle) in common, then the two triangles are the same. So we could answer Q2 as:

"Knowing the common angle and side, it is sufficient to show that the other adjacent side is congruent to prove that both triangles are congruent."

Q3: Knowing that *SUR* and *SUT* have in common the angle in *U* and the side *SU*, how can I prove that both triangles are congruent?

We know that it is sufficient to prove that the other side adjacent to the angle *U* is equal for both triangles, that is, $RU = UT$. So to prove that *SUR* and *SUT* are congruent triangles, it is sufficient to prove that $RU = UT$.

To prove that $RU = UT$, we have to prove that the distance from *U* to *R* is equal to the distance from *U* to *T*. We know about a perpendicular bisector has to do with distances being equal. Therefore, a perpendicular bisector of a segment AB is a line such that all its points are at the same distance to *A* and to *B*. So using this definition of perpendicular bisector, we could answer the previous question like this:

"We can prove that $RU = UT$ by proving that U is a point of a perpendicular bisector of RT."

Q4: How can I prove that *U* is in the perpendicular bisector of *R* and *T*? We already know that *SU* is the perpendicular bisector of *RT*, and so *U* is a perpendicular bisector of *RT*. Therefore, we have to prove that *U* is a perpendicular bisector of *RT*, and combining this fact with the answer to KQ4, we have proved that *U* is at the same distance of *R* and of *T*.

Now we could ask, "how can I prove the congruence of triangles?"

We observe that the triangles in question have one side and one angle in common. So we could ask what else must be proved to confirm that both triangles are congruent. Then it is sufficient to prove that a second side (adjacent to the angle) is equal in both triangles.

Then we could ask what we need to prove to confirm that *SUR* and *SUT* are congruent? Based on what we have said before, we could say that it is sufficient to prove that *RU = UT*.

Now we can try to prove that *RU = UT*

We know that a perpendicular bisector is a line such that all its points are the same distance from *R* and *T*. So one way to prove that *RU = UT*, that is, *U* is the same distance from R and from *T*, is to prove that *U* is in this perpendicular bisector. Now how can we prove that *U* is in the perpendicular bisector of *R* and *T*? And the answer comes from the question, because *SU* is the perpendicular bisector of *RT*, and so *U* is in it.

2.26 Prove that if triangle *RST* is equilateral and *SU* is a perpendicular bisector of *RT*, the area of triangle *RST* is $3^{1/2} (RS)^2 /4$.

To calculate the area of the triangle *RST*, we will calculate the area of the triangle *RSU* and the area of the triangle *TSU*, and add both of them.

Area of RST = Area of RSU + Area of TSU

Because *SU* is a perpendicular bisector of *RT*, we know the size of *SU* and we also know that both *RSU* and *TSU* are rectangular triangles; and it is easy to calculate the area of rectangular triangles.

Now the area of a rectangular triangle is easy to calculate. It is BASE*HEIGHT/2, where both BASE and HEIGHT are the sizes of the two perpendicular sides.

Therefore, we have

$$Area\ of\ RSU = SU*UR/2$$
$$Area\ of\ TSU = SU*UT/2$$

Now as SU is a perpendicular bisector of RT, then $RU = UT$, and so

$$RU = UT = TR/2$$

$$Area\ of\ RSU = SU*(TR/2)/2$$
$$Area\ of\ TSU = SU*(TR/2)/2$$

And also as RST is an equilateral triangle, we know that

$$TR = RS = ST$$

And so:

$$Area\ of\ RSU = SU*(RS/2)/2$$
$$Area\ of\ TSU = SU*(RS/2)/2$$

Then,

$$Area\ of\ RSU = Area\ of\ TSU = SU*(RS/2)/2$$

And so as $Area$ of $RST = Area$ of $RSU + Area$ of TSU, then we have that

$$Area\ of\ RST = 2*(Area\ of\ RSU) = 2*(SU*(RS/2))/2 = SU*RS/2$$

Now we have

$$Area\ of\ RST = SU*RS/2$$

We can see that SUT is a triangle such that the angle SUT is 90° and the angle STU is 60°, (because angle STU is the same angle as angle STR, that is 60°, because triangle STR is equilateral).

Then by definition of sin and cosine, we will have that

$$SU = ST*sin(< STU) = ST*sin(60^a) = ST * sqrt(3)/2$$

$$Area\ of\ RST = SU*RS/2 = (ST * sqrt(3)/2) * RS/2$$

And then,

$$Area\ of\ RST = ST * RS * sqrt(3) / 4$$

Now as $TR = RS = ST$ because RST is equilateral, we have that

$$Area\ of\ RST = (RS)^2 * sqrt(3) / 4$$

Chapter 3
Mathematical Terminology

3.2   For each of the following statements, obtain a new statement in the backward process by using a definition to answer the key question.

a. m is prime

With the backward process, we start from the conclusion and ask questions that will send us closer and closer to the first statement. For example, here we could ask:

Key Question: *How can I prove that m is prime?*

The definition of being prime is that p is prime if there is no n $(n > 1, n < p)$ such that n is a divisor for p. So we could answer the key question as:

I can prove that m is prime by proving that there is no
n (n > 1, n < m) such that n is a divisor for m

The next question or "previous step" in the backward process is the following:

KQ2: *How can I prove that there is no n (n > 1, n > m) such that*
n is a divisor for m?

b.   Triangle ABC is equilateral

With the backward process, we start from the conclusion and ask questions that will send us closer and closer to the first statement.

For example, we could ask the following key question:

How can I prove that ABC is equilateral?

Now what is the meaning of being *equilateral*?

ABC is equilateral if AB = AC = BC, that is, if its three sides are
equal

So we could answer the key question as:

I can prove that ABC is equilateral by proving that AB = AC = BC

Then the "previous step" in the backward process is the following:

KQ2: *How can I prove that in the triangle ABC, AB = AC = BC?*

c. p^2 is even

With the backward process, we start from the conclusion and ask questions that will send us closer and closer to the first statement.

Here we could ask the following question:

Key Question: *How can I prove that p^2 is even?*

Now what is the meaning of being *even*?

The definition of being even is that p^2 is even if 2 is a divisor for p^2.

So we could answer the question as:

Answer: *I can prove that **p^2** is even by proving that **2** is a divisor for **p^2**.*

Then the "previous step" in the backward process is the following:

KQ2: *How can I prove that 2 is a divisor for p^2?*

d. $(n)^{1/2}$ is rational

With the backward process, we start from the conclusion and ask questions that will send us closer and closer to the first statement.

Here we could ask the following question:

> **Key Question**: *How can I prove that $n^{1/2}$ is rational?*

Now what is the meaning of being *rational*?

The definition of being rational is that m is rational if there is p and q such that $m = p/q$

So we could answer the question as:

> **Answer**: *I can prove that $n^{(1/2)}$ is rational by proving that there are p and q integers such that $n^{(1/2)} = p/q$.*

Then the "previous step" in the backward process is the following:

> **KQ2**: *How can I prove that there exists p, q integers such that $n^{1/2} = p/q$?*

3.4   Use a definition to work forward from each of the following statements.

a.   For sets R, S, and T, $R = S$ union T

In the forward method, we start from the current statement to obtain new conclusions. In this case, the problem suggests using the definition to go forward. Therefore, we will use the definition of sets.

Definition of equality: *Two sets A and B are equal if and only if A is included in B, and B is included in A.*

In our example, we can substitute $R = S$ union T by its definition to obtain the next statement working forward.

> *For sets **R**, **S** and **T**, **R** is included in **S** union **T**, and **S** union **T** is included in **R**.*

b.   For functions f and g, the function $f + g$ is convex, where $f + g$ is the function whose value at any point x is $f(x) + g(x)$.

In the forward method, we start from the current statement to obtain new conclusions. In this case, the problem suggests using the definition to go forward. Therefore, what we will use is the definition of being convex.

Definition of convexity for functions:

F is a convex function, if and only if for all k (0 < k < 1) and for all s1, s2 in set S, we have

$$F(k*s1 + (1-k)*s2) \leq k * F(s1) + (1-k) * F(s2)$$

In our example, we will apply the definition of convexity to $f+g$.

"f + g is convex" means that "f + g" defined on a convex set X, and for all k (0 < k < 1) and for all x1, x2 in X,
*(f+g)(k.x1 + (1-k) * x2) ≤ k * (f+g)(x1) + (1-k) * (f+g)(x2).*

Now we obtain our next statement

For functions f and g, f+g is defined on a convex set X, and it is such that, for all x1, x2 in X, and for all k (0 < k < 1),
*(f+g)(k*x1 + (1-k *x2) ≤ k*(f+g)(x1) + (1-k)*(f+g)(x2)*

c.   For functions f and g and sets S and T, the function $f \geq g$ on (S intersection T).

The problem suggests using the definition to go forward. Therefore, what we will use is the definition of ($\geq$) for functions.

Definition of "≥" for functions is: $f \geq g$ in a set A if, for all x in A, f(x) ≥ g(x).

So in our example, we will apply this definition to obtain our new statement.

For functions f and g and sets S and T, f(x) ≥ g(x) for all x in (S intersection T).

3.8   Suppose that *A*, *B*, and *C* are statements. Is the statement "*A* implies (*B* OR *C*)" equivalent to the statement "(*A* AND NOT *B*) implies *C*"? Why or why not?

To see if they are the same statement, we will construct a table of values for both, and then observe if they will give the same result or not.

A	=>	(B	or	C)
T	**T**	T	T	T
T	**T**	T	T	F
T	**T**	F	T	T
T	**F**	F	F	F
F	**T**	T	T	T
F	**T**	T	T	F
F	**T**	F	T	T
F	**T**	F	F	F

The column, **=>**, means *implies*. We note that "*A* => *D*" is only *false* when *A* is *TRUE* and *D* is *FALSE*. In all other cases, we must put *true*.

(A	and	not	B)	=>	C
T	F	F	T	**F**	T
T	F	F	T	**T**	F
T	T	T	F	**T**	T
T	T	T	F	**T**	F
F	F	F	T	**F**	T
F	F	F	T	**T**	F
F	F	T	F	**F**	T
F	F	T	F	**T**	F

We note that "*A* => *D*" is only false when *C* is TRUE and *and* is FALSE (3 rows in the table). In all other cases, we must put *true*.

W can see that the results in both tables are different, and so "*(A and not B) => C*" is not the same as "*A* => *(B or C)*"

3.11   Use the *proposition* below to prove that if *a* and *b* are even integers then $(a+b)^2$ is an even integer.

Proposition: <u>If *n* is an even integer, then n^2 is an even integer</u>

We start from this statement

$$a \text{ and } b \text{ are even integers}$$

We then apply the definition of being *even* to go 1 step forward:

S1: *2 divides a and 2 divides b.*
There exists m, n integers such that a=2*m, b=2*n

Now we observe what happens with (a+b).

if a = 2*m and b = 2*n, then a+b = 2*m + 2*n = 2*(m+n)

Now we can write the next statement in the forward process.

S2: *there exists m, n integers such that (a+b) = 2*(m+n)*
Now observe that S2 implies that 2 is a divisor of (a+b).
And that (a+b) is even.

S3: *(a+b) is even*

Now we apply the property given by the problem.

If n is an even integer, then n^2 is an even integer.

And we obtain the following:

S4: *$(a+b)^2$ is even*

And we are done. We used the forward process to show that $(a+b)^2$ is even.

3.14   Prove if a and b are rational, then $a + b$ is rational.

We will use the forward method to prove the question. We will start from the following statement:

a and b are rational numbers

Now we apply the definition of "*being rational*" to go one step further.

Definition of being rational: *r is rational if and only if there exists p, q integers such that r = p/q.*

In our case, we have that a and b are rational, then

S1: *there exists p, q, m, n integers such that a = p/q and b = m/n*

Now let us look at what happens when we consider $(a + b)$

As $a = p/q$, and $b = m/n$, then we have

$$a+b = p/q + m/n = pn/qn + mq/nq = (pn + mq)/qn,$$

Now we write a new statement in the forward process.

S2: *there exists p, q, m, n integers such that*
$(a+b) = (pn + mq)/qn$

Now observe that *(pn + mq)* is an integer, and *qn* is also an integer.

Let us call $P = (pn + mq)$ *and* $Q = qn$. Then P and Q are integers, and so we have:

S3: *there exists P, Q integers such that (a+b) = P/Q*

And this is the definition of being (a+b) rational.

So we are done and we conclude that

S3: *(a+b) is rational.*

3.16 Suppose you already proved the proposition that, "If a and b are non-negative real numbers, then $(a+b)/2 \geq (ab)^{1/2}$".

a. Explain how we could use this proposition to prove that if a and b are real numbers satisfying the property that $b \geq 2|a|$, then $b \geq (b^2 - 4a^2)^{1/2}$

We proceed by the backward method. We start from the conclusion, and we want to try to find a sequence of statements that goes from the first one to the final one.

So we start with the following key question:

$KQ1$: How could I prove that $b \geq (b^2 - 4a^2)^{1/2}$?

Observe that we already know two things:

(1) The statement: $b \geq 2|a|$, and
(2) The property: $(a+b)/2 \geq (ab)^{1/2}$

We want to put the equation in the KQ1 in a way that resembles these two statements. First, observe that

$$(b^2 - 4a^2) = (b-2a) * (b+2a)$$

And so we can rewrite KQ1 as

$KQ2$: How could I prove that $b >= [(b - 2a)(b + 2a)]^{1/2}$?

Now we observe that if we call $A = (b - 2a)$, $B = (b + 2a)$, then $b = (A + B)/2$, and so we could answer KQ2 as thus:

To prove that $b \geq (b - 2a)(b + 2a)$, it is sufficient to prove that $(A + B)/2 \geq (A * B)^{1/2}$, where $A = (b - 2a)$ and $B = (b + 2a)$

So this sends us to the next KQ

$KQ3$: How could I prove that $(A + B)/2 \geq (A.B)^{1/2}$?
(where $A = b$-$2a$ and $B = b$+$2a$)

Now observe how similar a proposition we have.

The proposition states that

$$(a+b)/2 \geq (ab)^{1/2}, \text{ and we have } (A+B)/2 \geq (AB)^{1/2}$$

So to prove that $(A+B)/2 \geq (A*B)^{1/2}$ is based on the proposition, it is sufficient to prove that both A and B are non-negative.

Then to prove that $(A + B)/2 \geq A*B$, it is sufficient to prove that A and B are non-negative (where $A = b - 2a$ and $B = b + 2a$)

This sends us to KQ4

KQ4: *How could I prove that $A > 0$ and $B > 0$?*
(where $A = b-2a$ and $B = b+2a$)?

We know that $b \geq 2|a|$, as it is the initial statement provided by the problem. So b is positive (because $b \geq 2|a| \geq 0$, so $b \geq 0$).

Also as $b \geq 2|a|$, we have that $b \geq -2a$ and that $b \geq 2a$ (because one of these values is ≤ 0, and $\leq b$, while the other is equal to $2|a| \leq b$).

So from $b \geq 2|a|$, we get that $b \geq 0$, $b \geq 2a$, $b \geq -2a$, which means the following:

$(b - 2a) \geq 0$ and $(b + 2a) \geq 0$, that is, $A \geq 0$ and $B \geq 0$, that is, both A and B are nonnegative, as we wanted.

So to prove that A and B are nonnegative, it is sufficient to prove that $b \geq 2|a|$, and we know this is a true statement from the problem.

So we have proved that A and B are nonnegative, and we are done. We have proved the following:

A and B are non-negative. And so (as it was the answer to KQ4)
$(A + B)/2 \geq (AB)^{1/2}$ (where $A = (b-2a)$ and $B = (b+2a)$)

And by replacing A by $(b-2a)$ and B by $(b+2a)$ in the previous equation, we find that we have proved that

$$b \geq (b^2 - 4a^2)^{1/2}, \text{ as we wanted}$$

b. Use the foregoing proposition and part (a) to prove that if a and b are real numbers with $a < 0$ and $b \geq 2|a|$, then one of the roots of the equation $ax^2 + bx + a = 0$ is $\leq -b/a$.

Again with the forward process, we start from what we know and try to reach the desired result.

First, what are the roots of $ax^2 + bx + a = 0$?

Answer:

We know that the roots for an equation of degree 2 (general form, $ax^2 + bx + c = 0$) are given by the following formula:

$$\frac{-b - (b^2 - 4ac)^{1/2}}{2a} \quad \text{and} \quad \frac{-b + (b^2 - 4ac)^{1/2}}{2a}$$

In our case, $c = a$, so we have as roots:

$$\frac{-b - (b^2 - 4a^2)^{1/2}}{2a} \quad \text{and} \quad \frac{-b + (b^2 - 4a^2)^{1/2}}{2a}$$

Now as $b \geq 2|a|$, we know that $(b^2 - 4a^2) \geq 0$ (so the two roots exist).

And also as $b \geq 2|a|$, and by part (a), we know that

$$b \geq (b^2 - 4a^2)^{1/2}$$

Now by adding b to both sides, we obtain something more similar to what we want, a root compared to b.

$$2b \geq b + (b^2 - 4a^2)^{1/2}$$

Now dividing both sides by $2a$ we get

$$\frac{2b}{2a} \leq \frac{b + (b^2 - 4a^2)^{1/2}}{2a}$$

$$\frac{b}{a} \leq \frac{b + (b^2 - 4a^2)^{1/2}}{2a}$$

And multiplying both sides by -1

$$-b/a \geq [-b - (b^2 - 4a^2)^{1/2}] / 2a$$

This is as we wanted to prove

$$-b/a \geq \text{one of the two roots of the equation}$$

Chapter 4
The Construction Method

4.2   Suppose each statement in the following [except part (a)] is the conclusion of a proposition. Explain how we would apply the construction method to do the proof.

a.   At a party of n (≥ 2) people, at least two of the people have the same number of friends.

In the construction method, we construct a particular case that will prove the truth of the statement.

In this example, the construction is the following:

We look at all the people in the party. We consider all the possible pairs of people, and then we will pick a pair such that they have the same number of friends. Once we have this pair, we have proved the truth of the statement.

b.   The function $f(x)$ has an integer root.

In the construction method, we construct a particular case that will prove the truth of the statement.

We will try to find all the roots of the function $f(x)$, and then we will search for one root that is an *integer*. Once we have this integer root, then we have proved the truth of the statement.

c.   There is a point (x,y) where $x \geq 0$ and $y \geq 0$ that lies on the two lines $y = m_1 x + b_1$ and $y = m_2 x + b_2$.

In the construction method, we construct a particular case that will prove the truth of the statement.

In this example, we will search for the point of intersection of L1 and L2 (if they are non parallel), and verify that this point (x, y) is such that $x \geq 0, y \geq 0$. If L1 and L2 are parallel, we will look for one point at L1 that has $x \geq 0, y \geq 0$. In all cases, once we have this point, then we have proved the truth of the statement.

d.   Given an angle t, one can find an angle t' between 0 and pi whose tangent is larger than that of t.

In the construction method, we construct a particular case that will prove the truth of the statement.

In this example, the construction is the following:

Given t, with tangent tan(t), look for an angle in (0, pi) that has a larger tangent. Since the angle in (0, pi) that has the larger tangent is pi/2, then consider t' = pi/2. Then verify that for this value of t', it is true that tan(t') > tan(t). Once we have found the angle, then we have proved the truth of the statement.

e.   For the two integers a and b, at least one of which is not zero and whose greatest common divisor is c, there are integers m and n such that am + bn = c.

In the construction method, we construct a particular case that will prove the truth of the statement.

Here consider all the possible values for m and n, and look for a pair of values m and n that verify the wanted condition, that is, look for m and n such that $am + bn = c$. Once we find them, then we have proved the truth of the statement.

4.4   Would we use the construction method to prove each of the following propositions? Why or why not?

a.   The polynomial $x^{71} - 4x^{44} + 11x - 3$ has a real root.

No! We should not use the construction method here. Using the construction method would imply, that we have no method to find the root for a polynomial of degree 71. Therefore, we should solve the problem by reasoning using the properties of the polynomial, and NOT by calculating its roots.

b. If a and b are integers with a ≠ 0, for which a does not divide b, then there is no positive integer x such that $ax^2 + bx - a = 0$.

No! Here the construction method could be used to find a counter-example, but NOT to prove the assertion as it is, because the statement is a general one. Therefore, we can not build an example to show that

"there is no positive integer x such that $ax^2 + bx - a = 0$"

However, we could prove with the construction method the following:

"there is a positive integer x such that $ax^2 + bx - a = 0$"

It is sufficient to give one case to prove this last sentence.

c. If ABCD is a square whose sides have length s, then we can inscribe in ABCD a circle whose area is at least $3s^2 / 4$

Yes! Here we could use the construction method. We just have to explain how to construct the wanted circle. For example, this could be an explanation:

Construct the circle inscribed in ABCD such that it has diameter s. Then, the area of this circle is:

$$r^2 * pi = (s/2)^2 * pi = (s^2 / 4) * pi$$

And as pi > 3, we have that the area of the circle is greater than $(s^2/4) * 3 = 3s^2/ 4$. Therefore, we have constructed a circle inscribed on ABCD that has area > $3s^2 / 4$, as we wanted.

4.8   Prove that if *a* and *b* are integers *with a≠0,* and *x* is a positive integer such that $ax^2 + bx + b - a = 0$, then a/b.

We will proceed by the construction method. We want to prove that $a \, / \, b$, that is, there exists an integer, n, such that $a*n = b$. Therefore, in our construction, our goal is to construct such an *n*, that is, to construct the n such that $a*n = b$.

Now we know that *x* is positive and $ax^2 + bx + b - a = 0$, and so, $a(x^2 - 1) + b(x + 1) = 0$, *and as* $(x^2 - 1) = (x - 1)(x + 1)$, and we have that:

$$a(x - 1)(x + 1) + b(x + 1) = 0$$

And as $(x + 1) > 0$, because $x > 0$, we can divide both sides by $(x+1)$ *to* get the following:

$$a(x - 1) + b = 0$$
$$a(x - 1) = -b$$
$$a(1 - x) = b$$

Now as *x* is a positive integer, then $(1 - x)$ is also an integer, so let us call $n = (1 - x)$. Therefore, we have proved that $an = bn$ is an integer.

This means that we have completed the proof. We have constructed an *n* integer such that $a*n = b$, and we have proved that $a \, / \, b$.

70

4.10 Prove that if a, b, and c are integers for which a / (b+ c) and a/ b, then a / c.

We will proceed by the construction method. We want to prove that $a \:/\: c$, that is, there exists an n integer such that $a*n = b$. Therefore, in our construction, our goal is to construct such an n, that is, to construct the n such that $a*n = c$.

Now we know that there exists an m integer such that $a*m = (b+c)$ (because $a \:/\: (b+c)$), and that there exists a p integer such that $a*p = b$ (because $a \:/\: b$).

Now by subtracting the two equations, we find the following:

$$a*m - a*p = (b + c) - b, \text{ that is, } a*(m+p) = b + c - b = c$$

And so we have $a*(m+p) = c$.

So we have constructed the wanted n. As $m+p$ is an integer, we have found an integer n such that $a*n = c$, and this integer is $n = m+p$.

Therefore, we have proved that a / c.

4.12 Consider proving that if a, b, c , d and e are real numbers for which a certain property P holds, then the intersection of the two sets $S = \{(x,y) : y = ax^2 + bx + c\}$ and $T = \{(x,y) : y = dx + e\}$ is nonempty; that is, S intersection T ≠ empty set.

Find a property P that allows us to prove this proposition. Then prove the proposition.

We will proceed again by the construction method. We will try to find values a, b, c, d and e such that there is at least one point (x, y) in S and T.

First, we want to find values a, b, c, d and e such that both equations $y = ax^2 + bx + c$ and $y = dx + e$ are satisfied. Therefore, we need to find values of a, b, c, d and e such that $ax^2 + bx + c = dx + e$.

Consider, for example, the case $x = 1$. Then we need that $a + b + c = d + e$.

We will use this property for P:

Property P: $a + b + c = d + e$

If a, b, c, d and e verify this property, then the point $(1, a + b + c)$ is the same as the point $(1, d + e)$, and is both in S and T, so S intersection T is not empty. So if a, b, c, d and e satisfy the property P, then the proposition is proven, then

$$S \text{ intersection } T \text{ is not empty}$$

4.15 Explain what is wrong with the following condensed proof of the proposition that if m < n are consecutive integers and m is even, then 4 divides $m^2 + n^2 - 1$.

Proof: *Suppose that $n = m + 1$.*
The proof proceeds to note that for $k = m(m+1)$,
$m^2 + n^2 - 1 = 2k$.

There is a missing detail at the end of the proof. The correct proof should be:

Suppose that $n = m+1$. The proof follows by noting that for
$k = m(m+1)$, $m^2 + n^2 - 1 = 2k$.

Now as $k = m\,(m+1)$, then k is even (because either m or $(m+1)$ is even, and so $m(m+1)$ is even).

Therefore, $2k$ is a multiple of 4. As $m^2 + n^2 - 1 = 2k$, we have that $m^2 + n^2 - 1$, and that is a multiple of 4, as we wanted to prove.

Chapter 5
The Choose Method

In simple words, the "*Choose Method*" means the following:

Suppose that you want to prove that some property P holds for all elements from a set X. That is, you want to prove that "*for all x in X, P is true*". With the choose method, you will do two steps:

1) Choose an element y in X. y is ANY element chosen at random. You do not know what y to choose, you just know that y is in X.

2) Then prove that for this y, P is true.

5.2 Suppose you are trying to prove that, "If S and T are the sets defined by S = {(x,y) : x²+y² ≤ 16} and T = {(x,y) : 3x²+2y²≤ 125}, then for every element (x,y) of S, (x,y) is an element of T." Which of the following constitutes a correct application of the choose method?

Here we must prove that if (x, y) is an element of S, then *(x, y)* is an element of T, that is, S included in T.

We want to prove that *for all s in S, s is in T*. That is, comparing it with the general case "*for all x in X, P is true*". Here X is S and P is "*s is in T*".

Now by applying the choose method we would do the following:

1) Choose an element s in S. Again, s is ANY element of S. We just know that s *is in S*.

2) Then prove that for this s, s is in T.

a. Choose: A1 : real numbers x' and y'.
It must be shown that B1: (x',y') Element of T.

INCORRECT: This is not a correct application of the choose method, because A1 does not correspond to the general idea of picking an s in S. You are just considering two real numbers (x', y'), but not asking that (x', y') are in S, as it is required by the choose method.

b. Choose: A1: real numbers x' and y' with (x',y') Element of S.
It must be shown that B1: (x',y') is an element of X.

INCORRECT: At a first view, it may seem correct (comparing it with the correct version, that is, the number (5)). But, the error is that B1 should have said: (x', y') element of T, NOT of X.

c. Choose: A1: real numbers x' and y' with (x',y') Element of T.
It must be shown that B1: (x1,y') Element of S

INCORRECT: This is exactly the opposite of what we should have done. We should be proving that t included in s, not that s is included in t.

d. Choose: A1: real numbers x' and y', say 1 and 2, with $x^2 + y^2 = 5 \leq 16$ and therefore, (x',y') Element of T

INCORRECT: This one is tricky, because it seems to say all x, y such that $x^2 + y^2 \leq 16$ (if it said it, then it would be correct).

This is not correct because we need *(x, y) in S*, and not any other values.

e.   Choose: A1: real numbers x and y with (x,y)
Element of S.
It must be shown that B1: (x,y) Element
of T.

CORRECT: Corresponds to the two steps described previously
(that is, A1 corresponds to (1), and B1 to (2)).

5.4   Would you use the choose method to prove that for all
real numbers a, b, and c, if $4ac \leq b^2$, then ax^2+bx+c has
real roots? Why or why not?

No we would not. Because the choose method should be the
following:

1) A1: Consider three real numbers a, b, and c such that $4ac \leq b^2$

*2) Prove B1: for these 3 real numbers, $ax^2 + bx + c$ has real
roots.*

So we have not advanced at all. We are just saying the same
thing as before. We should find a different method.

5.6   Reword the following for-all statements in an equivalent
"If .., then ..." form.

a.   For every prime number p, p+7 is composite.

The choose method gives us the following two steps:

1) A1: Consider a prime number p.
2) Prove that *B1:* p+7 is composite.

And as an IF...THEN... sentence, we have the following:

> *If A1 then B1, that is, if p is a prime number, then
> p+7 is composite.*

b.   For all sets A, B, and C with the property A Subset B and B subset C, it follows that A subset C.

The choose method gives us the following two steps:

1) A1: Consider A, B, C such that A subset B and B subset C.
2) Prove that: B1: A subset C.

And as an IF...THEN... sentence, we have the following:

> *If A, B, C are such that A subset B and B subset C, then A subset C.*

c.   For all integers p and q with q ≠0, p/q is rational.

The choose method gives us the following two steps:

1) *A1*: Consider *p* and *q* rational, with *q* ≠ 0.
2) Prove that *B1*: *p/q* is rational.

And as an IF...THEN... sentence, we have the following:

> *If p, q are integers, with q ≠ 0, then p/q is rational.*

5.9   For the proposition and condensed proof given below, explain where (that is, in which sentence) why and how the choose method is used. Identify any mistakes you encounter.

Proposition*:*

> If $R = \{$*real numbers x:* $x^2 - x \geq 0\}$,
> $S = \{$*real numbers x:* $-(x-1)(x-3) \leq 0\}$, *and*
> $T = \{$*real numbers x:* $x \leq 1$ *or* $x \geq 3\}$,

> *Then R intersection S subset T.*

Proof*: To reach the conclusion, it will be shown that for all x Element R Intersection S, x Element of T. To that end, let x element of R intersection S.*

Because x is an element of R, $x^2 - x = x(x-1) \geq 0$*. Likewise, because x Element of S,* $-(x-1)(x-3) \leq 0$*. Combining these two yields that* $x \leq 1$ *or* $x \geq 3$.

The choose method is used in the sentence because "*to reach the conclusion, it will be shown that for all x element R intersection S, x element of T*".

So, the two steps are the following:

1) *A1: Consider x in R intersection S*
2) *Prove B1: x is in T.*

We want to prove that *for all x in R intersection S, x is in T.*

The given proof attempts to prove it. First it considers an *x element of R intersection S* by stating "*to that end, let x element of R intersection S.*"

Since x is in R, then x must satisfy, $x^2 - x \geq 0$, that is, $x(x-1) \geq 0$, because x is an element of R, $x^2 - x = x(x-1) \geq 0$.

Therefore, x must be in S, so $-(x-1)(x-3) \leq 0$, and because x is an element of S, $-(x-1)(x-3) \leq 0$).

Then by combining the two facts, we have the following:

$$x(x-1) \geq 0 \text{ implies that } x \leq 0 \text{ or } x \geq 1$$

(this conclusion, by studying the sign of this polynomial of 2nd degree).

$$-(x-1)(x-3) \leq 0 \text{ implies that } (x-1)(x-3) \geq 0$$
$$\text{and so } x \leq 1 \text{ or } x \geq 3$$

Therefore, the correct range where x satisfies both equations is the following:

$$x \leq 0 \text{ and } x \geq 3$$

This is not mentioned in the proof. Once you know that $x \leq 0$, you also know that $x \leq 1$, and so we have proved that $x \leq 1$ and $x \geq 3$.

Therefore, we have proved that x *is in T*, and this completes the proof.

5.12   Write an analysis that corresponds to the condensed proof given below. Indicate which techniques are used and how they are applied.   Fill in the details of any missing steps   where appropriate.

Proposition: *If p is a positive integer, then for all nonzero integers q and r having the same sign for which $q < r$, $p/q > p/r$.*

Proof: *Because q and r have the same sign, $p/(qr) > 0$. Now $r > q$, so multiplying both sides by p/qr results in $rp/(qr) > qp/(qr)$. It now follows that $p/q > p/r$, and so the proof is complete.*

1st sentence) *Because q and r have the same sign, $p/(qr) > 0$*

Here we observe that the proposition to be proved is, *if p is a positive integer, then for all nonzero integers q and r with the same sign, if $q < r$, then $p/q > p/r$*. Therefore, we start with the choose method and we get the following two steps:

1) A1: given p is a positive integer, consider q and r of the same sign such that $q < r$.

2) Prove that B1: $p/q > p/r$.

This is a step that is implied, but not written in the proof given. The proof starts with the forward method applied to A1.

Because A1 is true we know that p is positive and q and r have the same sign, and so qr is positive, and so $p/(qr)$ is positive, and so $p/(qr) > 0$.

Therefore, we use the choose method to start from *A1* and try to prove *B1*.

2nd sentence) *Now $r > q$ so multiplying both sides by p/qr results in $rp/(qr) > qp/(qr)$.*

Here we are continuing with the forward method applied to A1. We know from A1 that $r > q$.   Since $p/qr > 0$, then we can multiply both sides by p/qr to obtain $pr/qr > pq/qr$.

This is it. It is just one more step in the forward method.

3rd sentence) *now follows that p/q > p/r*

Here we just simplified. Since we have that $rp/qr > qp/qr$, then by simplifying, we get that $p/q > q/r$, as stated. It is just one more step in the forward method.

4th sentence) *and so the proof is complete*

Here we are doing a simple *backward step* based on B1 from the choose method. We wanted to prove that $p/q > q/r$, and so we are done.

5.14 Prove that for all real numbers a and b, at least on of which is not 0, $a^2 + ab + b^2 > 0$. Hint: use that fact that $a^2 + b^2 > (a^2 + b^2)/2$.

Using the choose method, we can do the following:

1) A1: consider a and b such that at least one is not zero.
2) Prove B1: $a^2 + ab + b^2 > 0$.

We will start from B1, and then use the backward method. The first question we should ask is the following:

KQ: *How could I prove that $a^2 + ab + b^2 > 0$?*

Answer: *Consider the **hint**, $a^2 + b^2 > (a^2 + b^2)/2$.*

Then, $a^2 + b^2 + ab > (a^2 + b^2)/2 + ab = (a^2 + b^2 + 2ab)/2$.

Now observe that $a^2 + b^2 + 2ab = (a+b)^2$. Therefore, we have found going forward from the hint that:

If $a^2 + b^2 > (a^2 + b^2)/2$, then $a^2 + b^2 + ab > (a + b)^2 / 2 > 0$, because $(a+b) \neq 0$ and so $(a+b)^2 > 0$.

Now to prove that $(a^2 + ab + b^2) > 0$, it is sufficient to prove that $a^2 + b^2 > (a^2 + b^2)/2$.

Now try to go forward from A_1 to reach the hint.

A_1) *a and b are such that at least one of them is not 0.*

In conclusion, $a^2 + b^2 > 0$, and as for all r positive, $r > r/2$, we then have that $a^2 + b^2 > (a^2 + b^2)/2$. Therefore, we have proved that

$$A_1 => the\ hint\quad (by\ forward\ process)$$
$$The\ hint => B_1\ (by\ backward\ process)$$

And so $A_1 => B_1$. Therefore, we have completed the proof. If a and b are such that at least one of them is not zero, then $a^2 + ab + b^2 > 0$.

5.15 A function f on one variable is strictly increasing if and only if, for all real numbers x and y with x < y, f(x) < f(y). Use the results from 5.14 to prove that the function f(x) = x³ is strictly increasing.

We want to prove that $f(x) = x^3$, that is, we want to prove that for all x, y that are real, if $x < y$, then $f(x) < f(y)$.

We can start with the choose method and these two steps:

1) *A1: x, y are two real numbers such that $x < y$*
2) *Prove B1: $x^3 < y^3$*

Let us start with B1, and work with the backward method.

The first question we should ask is:

KQ: *How could I prove that $x^3 < y^3$*

Answer: to prove that $x^3 < y^3$, it is sufficient to prove that
$$x^3 - y^3 < 0.$$

KQ2: *How could I prove that $x^3 - y^3 < 0$?*

Now observe that $x^3 - y^3 = (x - y)(x^2 + xy + y^2)$. Then it will be sufficient to prove that $(x - y)(x^2 + xy + y^2) < 0$.

Therefore, using KQ1 and KQ2 we get the following:

1) To prove that $x^3 < y^3$, it is sufficient to prove that $x^3 - y^3 < 0$.

2) To prove that $x^3 - y^3 < 0$, it is sufficient to prove that $(x-y)(x^2 + xy + y^2) < 0$.

Then to prove that $x^3 < y^3$, it is sufficient to prove that $(x - y)(x^2 + xy + y^2 < 0)$.

Now using *A1* let us work forward.

A1) x, y are such that $x < y$

First conclusion: $(x - y) < 0$

Second conclusion:

$(x - y)(x^2 + xy + y^2) < 0$ *if and only if* $(x^2 + xy + y^2) > 0$

Now combining this last conclusion from the forward method to the conclusion of the backward method, we get the following:

To prove that $x^3 < y^3$, it is sufficient to prove that $x^2 + xy + y^2 > 0$.

Now by problem 5.14, we know that this is always true for all x, y such that at least one of them is not 0.

Since $x < y$ (by A1), then x and y are different, and then at least one of them is not zero. Therefore, it is true that $x^2 + xy + y^2 > 0$. Then if *A1* holds, $x^2 + xy + y^2 > 0$, and this is sufficient to prove that $x^3 < y^3$.

Chapter 6
Specialization

6.4   For each of the following for-all statements, what conditions must the given object satisfy so that we apply specialization? Given that the object does satisfy those conditions, what can we conclude about the object?

a.

Statement: For all prime numbers p, p+7 is composite

Given object: An integer m.

When we have a statement such as the following:

"For all objects with a certain property, then something happens"

We have to apply *specialization*. We need that object to satisfy that *"certain"* property. And what we can conclude about this object is that *something happens*.

So in this particular case, we have *"for all prime numbers p, p+7 is composite"*. Then we can rewrite it as follows:

"For all numbers with the property of being a prime number, then p+7 is composite"

The property that the given object m must satisfy is being a prime number. And the conclusion that we would make is that $m + 7$ is composite.

b.

Statement:	For all elements x > 0 in a set S of real numbers, x is a root of the polynomial p(x).
Given object:	A real number y.

In the statement, *for all x real numbers with the property of x > 0 and x is an element of a set s,* then x is a root of the polynomial p(x).

Therefore, the property that y must satisfy is to be positive ($y > 0$), and to be an element of S.

Then the conclusion that we can make if y satisfies this property, is that y is a root of the polynomial $p(x)$.

c.

Statement:	Every triangle ABC with sides of length a = BC, b=CA, and c = AB, satisfies $c^2 = a^2 + b^2 - 2ab \cos(C)$.
Given object:	The isosceles right triangle ABC whose legs a = BC and b = CA are both equal to m.

In the statement, *for all triangles ABC (a = BC, b = CA, c = AB) of ANY KIND, then $c^2 = a^2 + b^2 - 2ab \cos(C)$,* we can see that there is no property required for the triangle to satisfy the conclusion.

Therefore, the property that the triangle *ABC* (given object) must satisfy is nothing.

Then the conclusion that we can make if the triangle, *ABC,* satisfies this property is that $c^2 = a^2 + b^2 - 2ab \cos(C)$.

d.   Statement: For all pairs of equilateral triangles ABC and DEF, if one side of triangle ABC parallel to one side of triangle DEF, then the other two sides triangle ABC are parallel to the corresponding sides of triangle DEF.

    Given object: The triangle CDE whose side DE is parallel to side DA of triangle FDA.

In the statement, *for all pairs of triangles ABC, DEF such that one side of ABC is parallel to one side of DEF, if the condition is* satisfied, then both triangles are equilateral.

Then we can conclude that the other two sides of triangles *ABC* are parallel to the corresponding sides of triangle *DEF*.

Therefore, the property that triangles *CDE* and *FDA* must satisfy is that *DE* is parallel to *DA*.

The conclusion that we can make if *CDE* and *FDA* satisfy this property, is that the other sides of *CDE* are parallel to the other sides of *FDA*.

6.6   To what specific object could we specialize each of the
following statements so that the result of specialization
leads to the desired conclusion? Verify that the object to
which we are applying specialization satisfies the
certain property in the for-all statement, so that we can
apply specialization.

a.   For-all statement: f is a function of one variable
such that for all real numbers
x, y, and t with $0 \le t \le 1$,
$f(tx+(1-t)y) \le tf(x) + (1-t) f(y)$.

In the statement, *for all real numbers x, y, t* with the property
where $0 \le t \le 1$, we can conclude the following:

$$f(tx + (1-t)y) \le tf(x) + (1-t)f(y)$$

We are looking for the conclusion where the function f satisfies
the following:

$$f(1/2) \le (f(0) + f(1))/2$$

Therefore, to specialize to the desired conclusion, we have to find
some values for *x, y, and t* such that:

1) They have the property where $0 \le t \le 1$

2) And $f(x + (1-t)y) \le tf(x) + (1-t)f(y)$

This will lead to the following conclusion:

$$f(1/2) \le (f(0) + f(1)) / 2$$

It is sufficient to take $x = 0$, $y = 1$, $t = \frac{1}{2}$, since these values
satisfy property (1), and applied to the general formula we get the
following:

$$f(0 + (1-1/2)*1) \le \frac{1}{2}$$
$$f(0) + (1 - \frac{1}{2}) * f(1),$$
$$f(0 + \frac{1}{2}) \le \frac{1}{2} * (f(0) + f(1))$$
$$f(1/2) \le (f(0) + f(1))/2 \quad \text{(as we wanted)}$$

b.   For-all statement: For all real numbers c and d for
which $c^2 \geq d^2$, $(c^2-d^2)^{1/2} \leq c$.

Here the statement is: For all REAL NUMBERS c, d with the
PROPERTY OF BEING $c^2 \geq d^2$, we can conclude $(c^2 - d^2)^{1/2} \leq c$

Desired conclusion: For the real numbers a,
$b \geq 0$, $(ab)^{1/2} \leq (a+b)/2$

To specialize to the desired conclusion, we have to find particular
values for c, d such that:

1) They satisfy the property: $c^2 \geq d^2$

*2) When applied to the general conclusion $(c^2 - d^2)^{1/2} \leq c$, will
lead to the following conclusion:*

For the real numbers a, b ≥ 0, $(ab)^{1/2} \leq (a+b)/2$.

It is sufficient to take $x = 0$, $y = 1$, $t = \frac{1}{2}$, since these values
satisfy property (1), and applied to the general formula we get the
following:

$$f(0 + (1\text{-}1/2) * 1) \leq \tfrac{1}{2}$$
$$f(0) + (1 - \tfrac{1}{2}) * f(1),$$
$$f(0 + \tfrac{1}{2}) \leq \tfrac{1}{2} * (f(0) + f(1))$$
$$f(1/2) \leq (f(0) + f(1))/2 \quad (as\ we\ wanted)$$

6.8   For real numbers u and v, prove that if u is an upper
bound for a set S of real numbers and $u \leq v$, then v is an
upper bound for S.

We must consider the definition of *upper bound*, which is the
following:

If x is an upper bound for S, this means that for all s in S, s $\leq$ x

We can see that this problem that can be solved by specialization,
since we can rewrite the previous definition as:

*For all x real numbers, if x $\geq$ s for all s in S, then
x is an upper bound of S.*

In our case, we want to prove that v is an *upper bound* of S. We want to prove that $v \geq s$ for all s in S.

Since we know that $v \geq u$, therefore this is an *upper bound of S*. And as u is an upper bound of S, then we know that $u \geq s$ for all s in S.

And since we know that $v \geq u$, we can conclude that $v \geq s$ for all s in S. This completes the proof, and we conclude that v *is an upper bound of S*.

6.10   For functions f and g of one variable, prove that if $g \geq f$ on the set of real numbers and x^* is a maximum of g, then for every real number x, $f(x) \leq g(x^*)$.

We know that $g \geq f$ implies the following:

For all x real number, $f(x) \leq g(x)$.

Also we know that x^* is a maximum for g, which means that

For all x real number, $g(x) \leq g(x^*)$.

Adding both yields that

For all x real number, $f(x) \leq g(x) \leq g(x^*)$, and So for all x real number, $f(x) \leq g(x^*)$.

6.12   Suppose that a,b, and c are real numbers. Prove that if $x^* = -b/(2a)$ is a maximum of the function $f(x) = ax^2 + bx + c$, then $a \leq 0$.

(Hint: Specialize x to a value of $x^* + E$, where e>0)

As $x^* = -b/2a$ is a maximum of the function $f(x) = ax^2 + bx + c$, then for all x, real numbers, $f(x) \leq f(x^*)$.

Now let us consider (following the hint) $x = x^* + E$ (where E > 0, any positive value).

It must be that $f(x) \leq f(x^*)$, that is, it must be that $f(x^* + E) \leq f(x^*)$, for all E > 0 close to zero.

And since we know that $f(x^* + E) = a(x^* + E)^2 + b(x^* + E) + c$, therefore our equation $f(x^* + E) \leq f(x^*)$ can be rewritten as:

$$a(x^* + E)^2 + b(x^* + E) + c \leq ax^{*2} + bx^* + c$$

$$ax^{*2} + 2ax^* E + aE^2 + bx^* + bE + c \leq ax^{*2} + bx^* + c$$

$$2ax^* E + aE^2 + bE \leq 0$$

Now dividing both sides by E, we get the following:

$$2ax^* + aE + b \leq 0$$

And remembering that $x^* = -b/(2a)$, we get the following:

$$2a(-b/2a) + aE + b <= 0$$

$$-b + aE + b \leq 0$$

$$aE \leq 0$$

And as $E > 0$, it must be that $a \leq 0$.

In addition, we know from the problem that a is not zero since from the definition, $x^* = -b/2a$, means we can not divide by zero.

6.14 Write an analysis of proof that corresponds to the condensed proof given below. Indicate which techniques are used and how they are applied. Fill in the details of any missing steps, where appropriate.

Proposition: *If f and g are convex functions, then f+g is a convex function.*

Then we know that f is convex, and g is convex.
We need to prove that f + g is also a convex function.

Proof: *To see that f + g is convex, let x, y, and t be real numbers with 0≤t≤1. Then, because f is convex, it follows that*

$$f(tx+(1-t)y) \le tf(x) + (1-t)f(y) \quad (i)$$

Likewise, because g is convex,

$$g(tx + (1-t)y) \le tg(x) + (1-t)g(y) \quad (ii)$$

Adding (i) and (ii) yields that

$$f(tx+(1-t)y) + g(tx+(1+(1-t)y) \le t[f(x)$$
$$= g(x)] + (1-t)[f(y) + g(y)]$$

Thus f + g is convex and the proof is complete.

We start with "the definition" of being convex. To be convex, a function must satisfy the following:

F is convex if for all x, y and t real numbers, with $0 \le t \le 1$, and it is true that $F(tx + (1-t)y) \le tF(x) + (1-t)f(y)$

This is why the proof states to *"let x, y and t be real numbers with $0 \le t \le 1$"*. We want to prove that for these values, the definition of $(f + g)$ being convex is satisfied.

To do so, we start with the known facts, that is, we know that f is convex, and this is translated (by the definition of convexity) in equation (i).

Likewise, because g is convex,
$$g(tx + (1-t)y) \le tg(x) + (1-t)g(y) \quad (ii)$$

Similarly, we use the definition of convexity (and the fact that g is convex) to formulate the equation (ii) for g.

Adding (i) and (ii) yields that

Now we must find a way to combine the well-known facts to obtain the goal of our proof. In this case it is simple. It is sufficient to add both equations which yield the following:

$$f(tx+(1-t)y) + g(tx+(1+(1-t)y)$$
$$\leq t[f(x) + g(x)] + (1-t)[f(y) + g(y)]$$

And this is a true statement. Now if we look back at the definition of convexity, this line means that the function given by $F(x) = f(x) + g(x)$ is convex.

An intermediate step that is missing, by definition of addition of functions, is the following:

$$(f + g)(x) = f(x) + g(x)$$

And so the function $F(x)$ is $(f + g)(x)$.

Then we have proved that F is convex. We have proved that f+g is a convex function.

Thus $f + g$ is convex and the proof is complete.

91

6.16   What, if anything, is wrong with the following proof?

Proposition: *If a, b, and c are real numbers with a <0 and x^* is a maximum of $f(x) = ax^2 + bx + c$, then for every real number $e>0$, $e \le (2ax^* + b)/a$.*

Proof: *Let e>0. It will be shown that $e \le (2ax^* + b)/a$. Because x^* is a maximum of f, by definition, for every real number x, $f(x^*) \ge f(x)$.*
In particular, for $x = x^* - e$, you have that:

$$a(x^*)^2 + bx^* + x \le a(x^* - e)^2 + b(x^* - e) + c$$
$$= a(x^*)^2 + bx^* + c - (b + 2ax^*)e + ae^2$$

Subtracting $a(x^*)^2 + bx^* + c$ from both sides and dividing by $e > 0$, it follows that

$$ae - (b + 2ax^*) \ge 0.$$

The desired result that $e \le (2ax^* + b)/a$ follows by adding $b + 2ax^*$ to both sides of the foregoing inequality and dividing by $a <0$.

The first line of the proof,

> *Let e>0. It will be shown that $e \le (2ax^* + b)/a$.*

This is correct. Our goal is then to prove that for any $e > 0$, $e \le (2ax^* + b)/a$.

Let us look at how the proof shows that $e \le (2ax^* + b)/a$.

> *Because x^* is a maximum of f, by definition, for every real number x, $f(x^*) \ge f(x)$.*

This is correct, although it should have said, "*by definition of maximum*", but the context is clear, therefore it is acceptable. The proof states, knowing that x^* is a maximum for f, then $f(x^*) \ge f(x)$ for all x (which is true by definition of maximum).

> *In particular, for $x = x^* - e$, we have that*

Then we will have that $f(x^*) \ge f(x)$, that is, $f(x^*) \ge f(x^* - e)$.

[Note that we just stated that $f(x^*) \geq f(x)$ for all x]

Therefore we have the following:

$$a(x^*)^2 + bx^* + x \leq a(x^* - e)^2 + b(x^* - e) + c$$
$$= \ a(x^*)^2 + bx^* + c - (b + 2ax^*)e + ae^2$$

There is an error. The first inequality is the translation of the fact that

$$f(x^*) \geq f(x^* - e)$$
$$a(x^*)^2 + bx^* + x \geq a(x^* - e)^2 + b(x^* - e) + c$$

However, it says "$\leq$", and not "$\geq$". Therefore, the inequality is wrong. Then each time we see a "$\leq$", we should replace it with "$\geq$", and replace "$\geq$" with "$\leq$".

Subtracting $a(x^*)^2 + bx^* + c$ from both sides and dividing by $e > 0$, it follows that $ae - (b + 2ax^*) \geq 0$.

This step is correct, since dividing by zero does not alter the equation. However, remember that there was an error in the previous step, and so the correct result should be the following:

$$ae - (b + 2ax^*) \leq 0$$

The desired result that $e \leq (2ax^* + b)/a$ by adding $b + 2ax^*$ to both sides of the foregoing inequality and dividing by $a < 0$.

If we add $(b + 2ax^*)$ to both sides, we get $ae \leq (b + 2ax^*)$, and then dividing both sides by a, we get the following:

$$e \geq (b + 2ax^*)/a.$$

Therefore, the proof was not correct. The error implied that the statement to be proved was false, that is, it is not true that $e \leq (b + 2ax^*)/a$, but it is the opposite, $e \geq (b + 2ax^*)/2a$.

Chapter 7
Nested Quantifiers

7.2   Rewrite each of the following statements, introducing appropriate notation and using nested quantifiers.

a.   A set S of real numbers has the property that, no matter which element is chosen in the set, you can find another element in the set that is strictly larger.

"A set S of real numbers has the property that" is just another way of stating the following:

   "There is a set S of real numbers that has the property that"

It is the same as:

   "There is a set S of real numbers that"

Therefore, rewriting the original statement yields the following:

"There is a set S of real number that, no matter which element is chosen in the set, you can find another element in the set that is strictly larger"

The second part of the statement, *"no matter which element is chosen in the set"*, is the same as:

   "for all elements in the set"

It is the same as:
   "for all s in S"

Therefore, rewriting the statement yields the following:

"There is a set S of real numbers such that, for all s in S, you can find another element in the set that is strictly larger."

The third statement,

"You can find another element in the set that is strictly larger"

Is another way of saying the following:

"There is another element in the set that is strictly larger", and
as *"another element"* means *"an element in S different than s"*,
that is, *"s' in S is different than s"*

And we can rewrite, *"there is an s' in S such that s' > s"*.

Adding all this yields the following statement:

*"There is a set S of real numbers such that, for all s in S,
there is an s' in S such that s' > s"*

b.   A function of a single variable has the property that for
some real number, the absolute value of the function is
always less that that number.

We can translate this statement, *"A function of a single variable
has the property that"* to the following:

*"There is a function of a single variable that has the property
that"*

It is the same as:

"There is a function of a single variable such that"...

*"There is a function of a single variable such that for some real
number, the absolute value of the function is always less that
that number"*

The statement, *"for some real number, the absolute value of the
function is always less than that number"* can be rewritten as:

*"There is some real number, such that the absolute value of the
function is always less than that number"*

And we can rewrite the statement as the following:

"There is a function of a single variable such that there is some real number, such that the absolute value of the function is always less than that number"

Now we observe that *"there is a function of a single variable"* can be rewritten better if we add a NAME to the function (we will use F, as the function name).

So we can rewrite it as

"there is a function F of a single variable such that"

Now adding the name of this function, *F,* to the rest of the statement, yields the following statement:

"there is a function F of a single variable such that there is some real number, such that the absolute value of the function f is always less than that number"

We can rewrite, *"The absolute value of the function F"* as just |F|.

"there is a function f of a single variable such that there is some real number, such that |F| is always less than that number"

Now we can give a number to the *"real number"* mentioned in the sentence. Let us call it *X.*

So we can rewrite the statement as thus:

"there is a function f of a single variable such that there is some real number X, such that |F| is always less than that number X"

Now we can substitute *"|F| is always less than that number X"* by just *"|F| < X",* and this yields to the following final statement:

"there is a function F of a single variable such that there is some real number X, such that | F | < X"

7.4   Are each of the following pairs of statements the same; that is, are these two statements both true?

a.   S1:   There is a real number $x \geq 2$ such that there is a real number $y \geq 1$ such that $x^2 + 2y^2 < 9$.

S2:   There is a real number $y \geq 1$ such that there is a real number $x \geq 2$ such that $x^2 + 2y^2 < 9$.

b.   S1:   There is a real number $0 \leq x \leq 1$ such that there is a real number $0 \leq y \leq 1$ such that $2x^2 + y^2 > 6$.

S2:   There is a real number $0 \leq y \leq 1$ such that there is a real number $0 \leq x \leq 1$ such that $2x^2 + y^2 > 6$.

c.   Based on your answers to part (a) and (b), when is the statement, "there is an object X, with a certain property P, such that there is an object Y, with a certain property Q such that something happens," the same as the statement, "there is an object Y with the property Q such that there is an object X with the property P such that something happens"?

CASE (**a**):   Yes they are the same.

We can see that the only difference is the order on the two *"there is"* parts of the sentence.

S1: *"there is an $x \geq 2$ such that there is a $y \geq 1$ such that..."*

and

S2: *"there is a $y \geq 1$ such that there is an $x \geq 2$ such that..."*

Both are the same, because if we consider an x that satisfies the first *"there is"* in s1, and a y that satisfies the second *"there is"* in s1, then we can use these same x, y to satisfy s2.

In addition, the values x, y that hold for *S2*, hold for *S1*, since both satisfy the *"there is"* parts.

So given any pair of values x, y, then "*S1 is true for x, y*," iff (if and only if) "*S2 is true for x, y*," and this means that they are equivalent.

98

CASE (**b**):

> Yes, they are the same. Again, if a pair (x, y) satisfies S_1, then they will satisfy S_2. So both S_1 and S_2 are equivalent.

CASE (c):

> Yes, they are the same. The properties P and Q do not interfere with one another. This is also the case in examples (a) and (b).

For example, in case (a), P is "$x \geq 2$" and Q is "$y \geq 1$", and as Q only relates to y (not to x), and P only relates to x (not to y), then they can be switched.

However, they will not be the same if for example, Q is dependent on x and y, as we can see in the following example:

S_1: *There is a real number $0 \leq x \leq 1$ such that there is a real number $0 \leq y \leq x$ such that...*

S_2: *There is a real number $0 \leq y \leq x$ such that there is a real number $0 \leq x \leq 1$ such that...*

As we can see, in S_1 the second condition $0 \leq y \leq x$ has the two variables (x and y) present. Therefore, if we invert the *there is* clauses, we obtain as S_2, an invalid statement, as it mentions x before defining it. It states that "*there is a real number y such that $0 \leq y \leq x$*", before having defined what x is.

7.6   Explain how to work forward from each of the following statements. How would you apply the proof techniques?

a.   For all objects X with a certain property P, there is an object Y with a certain property Q such that something happens.

STEP 1:

To work forward we apply specialization to the statement. Then based on the original statement, and given a particular X_o that has the certain property p, we can conclude that:

For X_o, there is an object Y with a certain property Q such that something happens.

STEP 2:

At this moment, we know that *"something happens"* for a certain Y with the property Q. Therefore, we could use this as part of a bigger proof where X_o is involved.

b.   There is an object X with a certain property P such that, for all objects Y with a certain property Q, something happens.

STEP 1:

Here we could apply specialization to the second part of the sentence to *"for all objects y with a certain property q"*, to obtain a statement valid for some y_o, that has the property q.

There is an object X with a certain property P such that, for Y_o, something happens.

STEP 2:

Again we know that *"something happens"* for a certain X with the property P (for our given Y_o). Therefore, we could use this as part of a bigger proof where Y_o is involved.

7.8   Prove that if S = {real numbers x > 0 : x^2 < 2}, then for every real number e > 0, there is an element x Element of S, such that x^2 > 2 -e.

In this problem we have to prove the following:

for all e > 0, there exists an x in S such that x^2 > 2 − e

For example, it is clear that such an *x* is *x* = *sqrt(2)*. The value verifies x^2 = *2* > *2 − e* for all positive values of *e*.

Since we have found the *x* such that x^2 > *2 − e* for all *e* > *0*, then we have completed the proof. Because to prove that something "*exists*", we just have to find it. Now you might ask, how do we find it?

We had to find a value in *S* that was greater than something; the first value considered was the sqrt (2) since it has a better chance of being greater than something.   By luck we tried this value first, and it worked. Therefore, the sqrt (2) verifies the desired condition that x^2 > *2 − e*. And so we are done.

7.10   Prove that if S = { (x,y) : x^2 + y^2 ≤1} and T= {(x,y): $(x-3)^2$ + $(y-4)^2$ ≥ 1 }, then there are real numbers a and b such that for every (x,y) Element of S, y ≤ ax + b and for every (x,y) Element of T, y ≥ ax+b.

Hint draw a picture of S and T and find an appropriate line y = ax + b.

We note that *S* is the circle with center *(0,0)* and a radius of *1*, and *T* is the circle with center *(3, 4)* and a radius of *1*.

The following equation is that of a circle with center (a, b) and a radius of R:

$$(x-a)^2 + (y-b)^2 \leq R^2$$

The problem asks us to find the values *a* and *b* such that for every (x, y) in S, *y* ≤ *ax* + *b*, and for every *(x, y)* in T, *y* ≥ *ax* + *b*.

Note that the equation $y \geq ax+b$ is the equation of one semi-plane (the semi-plane that is over the line of equation $y = ax+b$), and the equation $y \leq ax+b$ is the equation of the other semi-plane (the one that is under the given line).

So what we have to find is a *line* that passes between the circles S and T, without touching either one. Then we must find a way to obtain an S which is under the line and T which is over the line.

We must find the equation of a line that passes between them, line L.

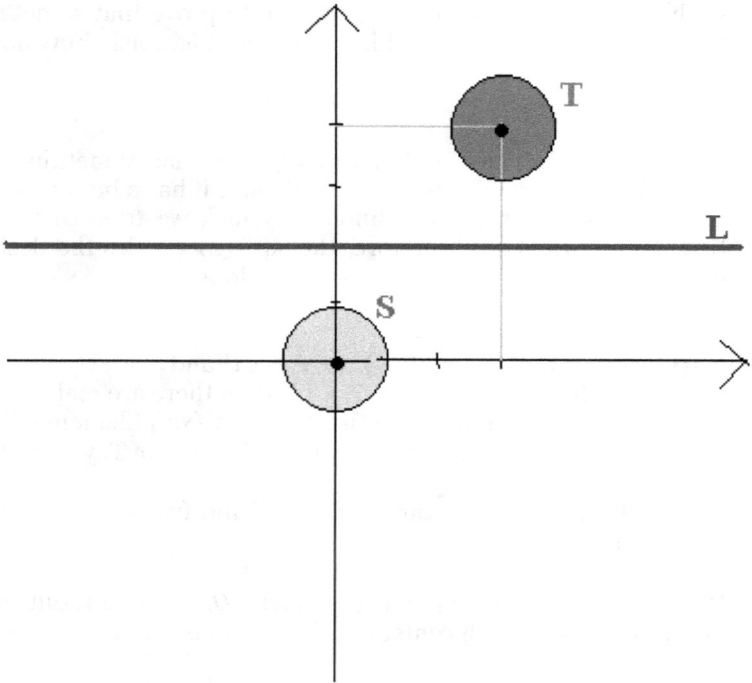

The figure above solves the problem, and we have used the line $y = 2$ as a solution.

Examine the following and you will see that *line y=2* does indeed satisfy all the conditions:

For all (x, y) in S, y ≤ 2
(as all the ordinates for points in S are ≤ 1).

For all (x, y) in T, y ≥ 2
(as all the coordinates for points in T are ≥ 3).

Therefore, we conclude that a possible line that satisfies the desired condition is $y = 2$. The case where $a = 0$ and $b = 2$ (the general equation is $y = ax + b$) .

7.12   Let f and g be functions of one variable. Prove that if f and g are onto (see Definition), then the function f o g is onto, where (f o g) (x) = f(g(x)).

Definition *onto*: a function *f* from the set of real numbers to the set of real numbers is *onto* if and only if for every real number y, there is a real number x such that $f(x) = y$.

We must prove that *f o g* is onto. We simply need to prove that *f o g* satisfies the definition of onto, that is, *f o g* is such that for every real number y, there is a real number x such that *f o g(x) = y*.

However, we know that for every real number y, there is a real number z such that *g(z) = y* [because *g* is onto]

In addition, we know that for every real number z, there is a real number x such that *f(x) = z* [because *f* is onto]

Adding both yields that "*for every real number y, there is a real number x such that f(g(x)) = y.*"

Now, as *f(g(x)) = f o g(x)*, we have proved that:

For every real number y, there is a real number x such that
$$fog(x) = y$$
Therefore,

f o g is onto
(because it satisfies the definition of being onto)

Chapter 8
Not Statements

The following list summarizes the rules for taking the NOT of statements that have a special form.

1. NOT [NOT A] becomes A

2. NOT [A AND B] becomes [(NOT A) OR (NOT B)]

3. NOT [A OR B] becomes [(NOT A) AND (NOT B)]

4. NOT [there is an object with a certain property, such that something happens, becomes "for all objects with the certain property, the something does not happen."]

5. Not [for all objects with a certain property, something happens] becomes "there is an object with the certain property such that the something does not happen."

8.2   Write the NOT of each of the following definitions in such a way that the word "not" doesn't appear.

a.   A sequence x_1, x_2, ... of real numbers is increasing if and only if for every integer k =1,2,...$x_k < x_{k+1}$

We will write NOT (statement), and then will apply rules *1 to 5* to obtain a final statement without the NOT.

A sequence x1, x2, ... of real numbers is **not** increasing is and only if **NOT** (for every integer k = 1, 2, ..., $x_k < x_{k+1}$)

NOT (for every integer k = 1, 2, ..., $x_k < x_{k+1}$), by rule 5, is the same as :

There is an integer k such that $x_k \geq x_{k+1}$

So the negative definition, *A sequence x1, x2, ... of real numbers is **not** increasing if **there is an integer k such that $x_k \geq x_{k+1}$***

b.   A sequence x_1, x_2, ... of real numbers is **decreasing** if and only if for every integer k=1,2,..., $x_k > x_{k+1}$

Again we have to compute **NOT** (for every integer k = 1, 2, ..., xk > xk+1), and by rule 5, this is the same as:

THERE IS an integer k such that $x_k <= x_{k+1}$

So the negative definition is, *A sequence x1, x2, ... of real numbers is **not** decreasing if **there is an integer k such that** $x_k \leq x_{k+1}$*

c.   A sequence x_1, x_2, ... of real numbers is **strictly monotone** if and only if the sequence is increasing or decreasing.

Now we have to compute **NOT** (the sequence is increasing or decreasing), by rule 3:

It is NOT increasing and it is NOT decreasing

And by applying the definitions for NOT being increasing and decreasing, we obtain that for (a) and (b):

*There exists an integer k such that xk $\geq$ xk+1
and there exists an integer k' such that $x_{k'} \leq x_{k'+1}$*

So the negative definition, *A sequence x1, x2, ... of real numbers is NOT **strictly monotone** if and only if there exists an integer k such that $x_k \geq x_{k+1}$ and there exists an integer k' such that $x_{k'} \leq x_{k'+1}$*

d.   An integer d is the **greatest common divisor** of the integers a and b if and only if (i) d/a and d/b and (ii) whenever c is an integer for which c/a and c/b, it follows that c/d.

Now we have to compute NOT (i) d/a and d/b and (ii) whenever c is an integer for which c/a and c/b), by rule 2:

NOT (d/a and d/b) OR NOT (whenever c is an integer for which c/a and c/b, it follows that c/d)

Now we observe that by rule 2 again, *NOT(d/a and d/b)* is equivalent to *NOT(d/a) OR NOT (d/b)*.

Then whenever *c* is an integer for which *c/a* and *c/b*, it follows that *c/d* is equivalent to *"for all integers c such that c/a and c/b, then c/d"*, then we must compute:

NOT (for all integers c such that c/a and c/b, then c/d), which is equivalent to:

> *THERE EXISTS an integer c such that c/a and c/b,*
> *and NOT (c/d).*

Therefore, the final statement is the following:

> *NOT(d/a) OR NOT(d/b) OR THERE EXISTS an integer c*
> *such that c/a and c/b, but NOT(c/d).*

It is not possible to eliminate these "NOT"s, as they are part of the statement. However, we can accept as a new definition the following:

An integer *d* is *NOT* the greatest common divisor of the integers *a and b,* if and only if, *d* does not divide *a*, or *d* does not divide *b*, or there exists an integer *c* such that *c/a* and *c/b*, but *c* does NOT divide *d*.

e.   The real number f' (x*) is the derivative of the function f at the point x* if and only if FOR ALL real number e > 0, there EXISTS a real number δ > 0 such that FOR ALL real number x with 0 < | x − x* | < δ, it follows that | f(x) − f(x*)|/|x − x*| < e

We are asked to find a NOT to the following definition:

NOT (FOR ALL real number e > 0, there EXISTS a real number
δ > 0 such that FOR ALL real number x with 0 < | x − x| < δ,*
it follows that |f(x) − f(x)|/|x − x*| < e)*

Since the sentence starts with "FOR ALL", we apply rule 5 and obtain:

THERE IS a real number e > 0 such that NOT (there EXISTS a real number δ > 0 such that FOR ALL real number x with $0 < |x - x^*| < δ$, it follows that $|f(x) - f(x^*)|/|x - x^*| < e$))

We apply rule 4, since it starts with a "there exists"; and we obtain:

THERE IS a real number e > 0 such that FOR ALL real numbers δ > 0 it happens that NOT (FOR ALL real number x with $0 < |x - x^*| < δ$, it follows that $|f(x) - f(x^*)|/|x - x^*| < e$).

We apply rule 5, and then we can write the following statement:

THERE IS a real number e > 0 such that FOR ALL real numbers δ > 0, THERE EXISTS a real number x with $0 < |x - x^*| < δ$, such that NOT($|f(x) - f(x^*)|/|x - x^*| < e$).

Then,

$$NOT(|f(x) - f(x^*)|/|x - x^*| < e) \text{ is } |f(x) - f(x^*)|/|x - x^*| \geq e$$

So the final statement is:

THERE IS a real number e > 0 such that FOR ALL real numbers δ > 0, THERE EXISTS a real number x with $0 < |x - x^*| < δ$, such that $|f(x) - f(x^*)|/|x - x^*| \geq e$.

Therefore, we can now write the following opposite definition:

The real number f'(x*) is NOT the derivative of the function f at the point x* if and only if THERE IS a real number e > 0 such that FOR ALL real number δ > 0, THERE EXISTS a real number x with $0 < |x - x^*| < δ$, such that $|f(x) - f(x^*)|/|x - x^*| \geq e$.

8.4 Write the negation of the conclusion in each of the following implications. State your answer in such a way that the words "no" and "not" don't appear.

a. If a is a real number, then there is a real number x such that $x = a^{-x}$

The negation of the conclusion is:

NOT(there is a real number x such that $x = a^{-x}$).

And by applying rule 5 (as it is a "THERE IS" rule):

For all real number x, NOT($x = a^{-x}$).

For example,

For all real number x, $x \neq a^{-x}$ (where "$\neq$" means "not equal").

b. A implies NOT B

Here the conclusion is *NOT B*, so we must compute:

NOT (NOT B) = B (by rule 1)

So the answer is **B.**

c. A implies (B implies C)

Here the conclusion is (*B implies C*), so we must compute:

NOT (B implies C).

Now we use the fact that the statement "*B implies C*" is equivalent to "*NOT B or C*". So we need to compute:

NOT (NOT B or C).

By applying rule 3, the previous sentence is equivalent to:

NOT (NOT B) AND NOT(C).

This in turn is equivalent (by rule 1) to:

B AND NOT(C).

Therefore, the opposite of the original conclusion is:

B AND NOT(C).

d. If u is a least upper bound for a set S of real numbers, then there is real number e>0, and there is an element x Element of S such that X > u -e.

Here the conclusion is:

"FOR ALL real numbers e>0, THERE IS an element x Element of S such that X > u -e."

So the conclusion is:

"For all real numbesr e > 0, THERE IS an element x of S such that X > u − e".

So we must compute the NOT of the conclusion:

NOT (For all real number e > 0, THERE IS an element x of S such that X > u − e).

Now applying rule 4, this is the same as:

THERE IS a real number e > 0 such that NOT (THERE IS an element x of S such X > u − e).

And applying rule 5, we get the following:

THERE IS a real number e > 0 such that FOR ALL element x of S, NOT(X > u − e).

And by using the fact that *NOT(X > u − e)* is the same as "*X ≤ u − e*", then we obtain the following:

THERE IS a real number e > 0 such that FOR ALL element x of S, X ≤ u − e.

8.6   For each of the following statements A implies B, what statement(s) will you work forward from and what statement(s) will you work backward from it if you want to prove that NOT B implies NOT A?

a.   A implies (C AND D)

We want to prove that *NOT B* implies *NOT A*, and as *B = (C AND D):*

> *NOT (C AND D) implies NOT A.*

We want to prove that:

> *(NOT C OR NOT D) implies NOT A.*

We used rule 2 to translate

> *NOT (C AND D)* to *(NOT C OR NOT D).*

Now we work forward from the statements *NOT C or NOT D*, to try to prove *NOT A*; or we can work backward from the statement *NOT A*, and try to prove that *(NOT C OR NOT D)* is necessary to prove *NOT A*.

b.   A implies (C OR D)

We want to prove that *NOT B* implies *NOT A*, and *B = (C OR D):*

> *NOT (C OR D) implies NOT A.*

We want to prove that:

> *(NOT C AND NOT D) implies NOT A.*

We used rule 2 to translate

> *NOT (C OR D) to (NOT C AND NOT D).*

So we will work forward from the statements *NOT C* and *NOT D*, to try to prove *NOT A*, or we will work backward from statement *NOT A*, and try to prove that *(NOT C AND NOT D)* is necessary to prove *NOT A*.

111

d. Suppose that m and n are integers. If either mn is divisible by 4 or n is not divisible by 4, then n is an even integer and m is an odd integer.

We want to prove that *NOT B* implies *NOT A*, and as
A = "*If either mn is divisible by 4 or n is not divisible by 4*" and
B = "*n is an even integer and m is an odd integer*", then

$\qquad$ *NOT (n is an even integer and m is an odd integer) .*

And this implies:

$\qquad$ *NOT (either mn is divisible by 4 or n not divisible by 4).*

Then we want to prove the following:

$\qquad$ *(n is not an even integer OR m is not an odd integer).*

And this implies:

$\qquad$ *NOT (either mn is divisible by 4 or n not divisible by 4)*

So we will work forward from statements **n is not an even integer** or **m is not an odd integer**, to try to prove NOT(either *mn* is divisible by 4 or *n* not divisible by 4).

We can also work backward from statement NOT(either *mn* is divisible by 4 or not divisible by 4) to try to prove that at least one of (*n* is not an even integer) or (*m* is not an odd integer) is necessary to prove the conclusion.

8.8 Provide a counterexample to show that each of the following statements is not true.

a. For every real number x>0, $(x)^{1/2} \leq x$

We must prove that NOT (For every real number $x > 0$, $x^{1/2} \leq x$).

$\qquad$ *THERE IS a real number $x > 0$ such that $x^{1/2} > x$.*

112

A counter example is any number less than 1, for example ½:

In the case of 1/2, we have:

$$\tfrac{1}{2}^{1/2} = 1/sqrt(2) > \tfrac{1}{2}, \text{ because } sqrt(2) < \tfrac{1}{2}$$

So we have proved that for $x = \tfrac{1}{2}$, it is true that $x^{1/2} > x$, and that is a contradiction of the original statement.

b.   For every positive integer n, $n^2 + n + 41$ is prime.

A counter example, we will try to find a positive integer such that $n^2 + n + 41$ is NOT prime.

We can try 1, 2, 3, 4, etc.

$1^2 + 1 + 41 = 43$	(prime – so it is not useful)
$2^2 + 2 + 41 = 47$	(prime – so it is not useful)
$3^2 + 3 + 41 = 53$	(prime – not useful)
$4^2 + 4 + 41 = 57$	MULTIPLE OF 3, so not prime

So we have found a counter example that if $n = 3$, then it is NOT TRUE that $n^2 + n + 41$ is prime.

c.   If x is a real number, then $x \neq 2^{-x}$ (HINT: use trial and error on the interval [0,1] to find an approximate value for x.)

A counter example would be, if we have to find a real number x such that
$x = 2^{-x}$.

Let us try to use the hint. We observe that x is an increasing function and 2^{-x} is a decreasing function.

We can see that when $x = 0$, $2^{-x} = 1$, and that when $x = 1$, $2^{-x} = \tfrac{1}{2}$.

So at $x = 0$, $x - 2^{-x}$ is negative, and at $x = 1$, $x - 2^{-x}$ is positive. And as the function is continuous, there must be at least one intermediate point in the interval [0, 1] such that
$x = 2^{-x}$.

This completes the proof that the statement $x \neq 2^{-x}$ is false, but we will try to find a counter-example.

We have the following:

$$At\ x = 0, x - 2^{-x} < 0$$

$$At\ x = 1, x - 2^{-x} > 0$$

We can see what happens at $x = \frac{1}{2}$:

$$x - 2^{-x} = \frac{1}{2} - 2^{-1/2} = \frac{1}{2} - \frac{1}{2^{1/2}} = (sqrt\ 2 - 2)/2\ sqrt(2) < 0$$

So x must be between $\frac{1}{2}$ and 1.

We can see what happens at $x = 3/4$:

$$x - 2^{-x} = 3/4 - 2^{-3/4} = 3/4 - \frac{1}{2^{3/4}} = (3.2^{3/4} - 4)/(4.2^{3/4} > 0$$

So x must be between $\frac{1}{2}$ and $\frac{3}{4}$.

Let us look at what happens at $x = 5/8$:

$$x - 2^{-x} < 0.$$

So x must be between $5/8$ and $\frac{3}{4}$.

Let us look at what happens at $x = 11/16$:

$$x - 2^{-x} < 0.$$

So x must be between $11/16$ and $3/4$

Let us look at what happens at $x = 23/32$:

$$x - 2^{-x} > 0$$

So x must be between $11/16$ and $23/32$.

We can continue using this method until we get any precision that we want.

Chapter 9
The Contradiction Method

9.2   Consider applying contradiction to show that if a and b are integers and b is odd, then ± 1 are not roots of $ax^2 + bx + a$.

a.   What statement(s) would you work backward from?

To use contradiction to prove that $S1$ implies $S2$, then prove that $NOT (S2)$ implies $NOT (S1)$.

In our case, contradiction can be used to prove the following:

"If a and b are integers and b is odd, then $+1$ and -1 are not roots of $ax^2 + bx + a$".

To proceed, we must prove that

> NOT ($+1$ and -1 are not roots of $ax^2 + bx + a$)
> $IMPLIES$ NOT (a and b are integers and b is odd).

We can rewrite it as

> *At least one of $+1$ and -1 IS a root of $ax^2 + bx + a$*
> *IMPLIES NOT(a and b are integers) OR NOT (b is odd)*

If at least one of $+1$ or -1 IS a root of $(ax^2 + bx + a)$ THEN
(NOT (a is integer) or NOT (b is integer) OR b is even).

If at least one of $+1$ or -1 IS a root of $(ax^2 + bx + a)$ THEN
(a is not integer or b is not integer or b is even).

Now working backward from this statement, a is not an integer, or b is not an integer, or b is even.

115

b. At the end of the proof, a mathematics student said, "...and because I have been able to show that b is even, the proof is complete." Do you agree with the student? Why or why not?

Yes the proof is complete if he was able to prove that b is even. Then he have proved that the following is true:

"a is not integer or b is not integer or b is even"

However, it is *not* possible to prove that b is even in *all* cases. Therefore, the proof is wrong. However, if we agree that he was able to prove that b is even, then yes indeed, he has proved that *"a is not integer or b is not integer or b is even"*, and so the proof is correctly done.

9.4 Reword each of the following statements so that the word "not" does not appear.

a. The real number ad - bc is not equal to 0.

Instead of saying *"not equal to 0"*, we can say *"different than 0"*.

The real number ad – bc is different than 0

b. The triangle ABC is not equilateral.

Instead of saying *"not equilateral"*, use the definition of being equilateral.

The triangle ABC is NOT (such that all its sides are equal)

And the following statement is true:

"The triangle ABC is such that at least one of its sides is different in size from other side"

<u>c.</u>   The polynomial $a_0 + a_1 x + ... + a_n x^n$ has no real root.

Instead of saying *"has no real root"*, use the definition of polynomial having real root.

*"**has real root**" = "there exists x real such that P(x) = 0, where P is the polynomial".*

So we can rewrite the previous sentence as:

The polynomial $a_0 + a_1 x + ... + a_n x^n$ is such that NOT (there exists x real such that P(x) = 0, where P is the polynomial)

Now from rule 5 for NOT,

NOT (THERE IS x such that something) = FOR ALL(x NOT (something)).

"NOT (there is x real such that P(x)=0)" = FOR ALL(x real, P(x) different than 0).

Therefore, we should rewrite the sentence as:

The polynomial $a_0 + a_1 x + ... + a_n x^n$ is such that FOR ALL (x real, P(x) different than 0, where $P(x) = a_0 + a_1 x + ... + a_n x^n$).

Then, going one step further, we can write it as:

The polynomial $a_0 + a_1 x + ... + a_n x^n$ is such that FOR ALL x real, $a_0 + a_1 x + ... + a_n x^n$ is different than 0

9.6   When trying to prove each of the following statements, which techniques would you use and in which order? Specifically, state what you would assume and what you would try to conclude. (Throughout, S is a given set of real numbers and all of the variables refer to real numbers).

This proof requires using *ALL* the *knowledge* we have learned in chapters 1 to 8. In each chapter we have learned a different technique applicable to a different kind of problem. Let us recall briefly the names of these techniques and how they work, and when they can be applied. Then we can solve this proof.

Chapter 1
Proofs and Answers

This method proof is simple. We will not use it often.

Chapter 2 and 3
The Forward-Backward Method

We will use this method often.

The forward process and the backward process work as reverses of each other.

To prove that "*A implies B*", the forward method starts supposing that A is true, and deduce from it a conclusion that is an intermediate step (A') between A and B. Then, we must prove that "*A' implies B*" to complete the proof.

To prove that "*A implies B*", the backward method starts from B (the conclusion), and asks a key question, for example:

How can I prove that B is true?

We obtain an intermediate statement, B', such that "*B' implies B*" is true. The method is useful if B' is intermediate between A and B, because the further step is to prove that "*A implies B'*"

Chapter 4:
The Construction Method

This is useful to prove statements of the kind

"THERE IS x with a certain property".

We construct the desired x (build the x that has the desired property, to prove that x exists).

Chapter 5:
The *Choose Method*

This is useful to prove statements of the kind

"*FOR ALL x with a certain property, then something happens*".

We choose an element with the certain property, and then must prove that, by the fact of having this property, the "*something happens*" must be true.

Chapter 6:
Specialization

This is useful to prove that for a certain object, something happens. Suppose that you already know that

"*for all x, if x has certain property, something happens*".

We just have to prove that our object verifies the "*certain property*".

Chapter 7 and 8
Nested Quantifiers and Not Statements

This deals with "*for all*" and "*there is*" statements. The main point is to be able to identify *the object, the certain property,* and *the something that happens,* in each sub-statement starting with "*for all*" or "*there is*".

Chapter 9:
The *Contradiction Method*

This method helps us to show how to use the fact that "*A implies B*" and "*NOT B implies NOT A*" are equivalent. In certain cases, it is easy to prove that *NOT B implies NOT A,* using the contradiction method.

a.   For every real number e>0, there is an element x Element of S such that x > u - e (where u is a given real number).

We have seen in Chapter 7 how to translate this sentence as a set of two nested *"for all* and *there is"*:

FOR ALL e real number, e > 0, THERE IS an element x of S such that x > u − e (where u is a given real number)

To prove it, we apply the choose method (to prove *"for all"* statements):

> *Given e real number, e > 0, THERE IS an element x of S such that x > u − e (where u is a given real number).*

Now we have to prove the *THERE IS* sentence for the given real number, *e > 0.*

How do we proceed? As stated in Chapter 4, one way to prove the "*THERE IS*" statement is to use the construction method. Then we can do the following:

> *Given a real number, e > 0, we can CONSTRUCT an element x of S such that x > u − e.*

Then we have completed the proof. If it is possible to construct this element *x* (for example, to find an element in *S* such that $x > u − e$), we would be proving the original statement.

b.   There is a real number y > 0 such that for every element x Element of S, f (x) < y (where F is a function of one variable).

Here we use Chapter 7 lessons to build *THERE IS* and *FOR ALL* −statements:

"THERE IS a real number y > o such that FOR ALL x element of S, f(x) < y (where f is a function of one variable)".

To prove *THERE IS* statements, the construction method is used.

We want to construct a real number $y > 0$, such that *FOR ALL* x element of $S, f(x) < y$.

We want to find a real number $y > 0$ such that *FOR ALL* x element of $S, f(x) < y$.

Then we can do the construction depending on S and F.

<u>c.</u> For every line l in the plane that is parallel to, but different from, l', there is no point l that is also in l' (where l' is a given line in the plane).

We can use Chapter 7 to build a *THERE IS* and a *FOR ALL* statement.

"FOR ALL line l in the plane such that l is parallel to l' and l is different to l', THERE IS no point on l that is also in l' (where l' is a given line in the plane)".

Now using Chapter 5, we can use the *choose method* to prove the *FOR ALL* statement.

"Given line l in the plane such that l is parallel to l' and is different to l', THERE IS no point on l that is also in l' (where l' is given line in the plane)".

As the conclusion is *"negative"*, we could try to apply the principles from Chapter 8, and apply the *contradiction method* to prove this statement. Instead of proving *A implies B*, we could prove *NOT B* implies *NOT A*:

Given line l in the plane, NOT (THERE IS no point on l that is also in l' (where l' is given line in the plane) IMPLIES NOT(l is parallel to l' and different to l').

We can rewrite it as:

Given lines l, l' in the plane, NOT (THERE IS no point on l that is also in l') IMPLIES NOT (l is parallel to l' and different to l')

Now as (*THERE IS NOT(x) that something*) is the same thing as *FOR ALL x, NOT (something)*, we have that our statement is:

Given lines l, l' in the plane, NOT (FOR ALL point on l, the point is NOT in l') IMPLIES NOT (l is parallel to l' and different to l')

Remember that NOT(FOR ALL x, something) = THERE EXISTS (x such that NOT (something):

Given lines l, l' in the plane, THERE EXISTS a point on l such that NOT(the point is NOT in l') IMPLIES NOT (l is parallel to l' and different to l)

Now we observe that

$$NOT(l \text{ is } NOT \text{ in } l') = n \text{ } l') = \text{the point is in } l'$$

And

$$NOT(A \text{ and } B) =$$
$$NOT A \text{ or } NOT B, \text{ so}$$
$$NOT (l \text{ is parallel to } l' \text{ and different to } l') =$$
$$NOT (l \text{ is parallel to } l') \text{ OR}$$
$$NOT (l \text{ is different to } l')$$

Therefore, the statement can be rewritten as:

Given lines l, l' in the plane, THERE EXISTS a point on l such that the point is also in l' IMPLIES (NOT (l is parallel to l') OR NOT(l is different to l')).

122

Now given lines l, l' in the plane, *THERE EXISTS* a point on l such that the point is also in l' IMPLIES (l is not parallel to l', or l = l').

For example,

> *Given lines l, l' in the plane, THERE EXISTS a point on l intersection l' IMPLIES (l is not parallel to l' OR l = l')*

Now to prove this statement, we can just go forward using the definition of *parallelism* and *identity of lines*.

Given any 2 lines, they can be

> *1) parallel and not intersecting*
> *2) parallel and identified*
> *3) non-parallel*

Cases (1) and (3) are when the lines have an intersection, and case (2) is a case when the lines DO NOT have an intersection.

So by the hypothesis we know that the intersection EXISTS, and going forward we know that the lines may only be of (1) or (3), therefore implying that the conclusion is TRUE.

9.8 Prove by contradiction, that there do not exist positive real numbers x and y, with x ≠ y, such that $x^3 - y^3 = 0$.

We want to prove the following:

> *There do not exist positive real numbers x and y with x ≠ y such that $x^3 - y^3 = 0$.*

To prove "*A implies B*" by contradiction, we must prove (*NOT B implies NOT A*), that is equivalent to "*A implies B*"

In our case, we can translate our statement to the "*A implies B*" as thus:

> *Given x and y real numbers such that x ≠ y, THEN it is impossible that $x^3 - y^3 = 0$.*

Now, rewriting it in a friendlier statement:

FOR ALL x, y real numbers such that x ≠ y, NOT($x^3 - y^3 = 0$).

For example,

Given x, y real numbers: x ≠ y IMPLIES NOT($x^3 - y^3 = 0$).

Now to prove this by contradiction, we invert the "*A implies B*" sentence to a "*NOT B implies NOT A*" sentence, and we get the following:

Given x, y real numbers: NOT (NOT($x^3 - y^3 = 0$) IMPLIES NOT (x ≠ y).

For example,

Given x, y real numbers: ($x^3 - y^3 = 0$) IMPLIES (x = y).

It is sufficient to prove the original statement by contradiction. Now we start from the hypothesis:

$$x^3 - y^3 = 0$$

Then we go forward. The first conclusion we obtain using the fact that $x^3 - y^3 = (x - y)(x^2 + xy + y^2)$ is:

(x − y)(x² + xy + y²) = 0 (because $x^3 - y^3 = (x - y)(x^2 + xy + y^2)$

And now using the fact that $x^2 + xy + y^2$ is NEVER equal to ZERO, (because $x^2 + xy + y^2$ has NO real roots), we then conclude:

$$(x - y) = 0$$

Therefore, x = y as we wanted.

9.10 Suppose that a, b, and c are real numbers with $c \neq 0$. Prove, by contradiction that if $cx^2 + bx + a$ has no rational root, then $ax^2 + bx + c$ has no rational root.

Again, to prove this by contradiction, we have to translate a sentence "*A implies B*" to its version "*NOT B implies NOT A*", and then prove it.

To start, put the original sentence as "*A implies B*". Let us suppose that *a*, *b* and *c* are real numbers with $c \neq 0$. Then prove by contradiction that:

IF $cx^2 + bx + a$ has NO rational roots, THEN $ax^2 + bx + c$ has NO rational root.

To prove it by contradiction, we have to prove the following:

IF NOT($ax^2 + bx + c$ has NO rational root) THEN NOT ($cx^2 + bx + a$ has NO rational root).

Which is the same as:

IF $ax^2 + bx + c$ has rational root, then $cx^2 + bx + a$ has rational root

To prove it this, we can use the forward method. We start from the first statement, and try to prove the conclusion.

We start from $(ax^2 + bx + c)$ has a rational root. There is p rational such that $ap^2 + bp + c = 0$.

We know that $p \neq 0$, because $c \neq 0$, and so 0 is not a root of $ax^2 + bx + c$. Therefore, we can divide the previous expression by p^2, and obtain:

$$a + b/p + c/p^2 = 0$$

$$a + b(1/p) + c(1/p)^2 = 0$$

Then, $(1/p)$ is a root of $a + bx + cx^2$

And so $a + bx + cx^2$ has a rational root (because $1/p$ is rational, as p was rational; and $1/p$ is a root of $a + bx + cx^2$).

9.12   Prove, by contradiction, that if p and q are integers with
p ≠ q and p is prime and divides q, then q is not prime.

To prove this by contradiction, we have to translate "*A implies B*"
to "*NOT B implies NOT A*", and then translate it.

Let us start by putting the sentence in the form "*A implies B*":

*IF p and q are integers with p ≠ q and p is primer and divides q,
THEN q is not prime.*

In fact, the sentence is better expressed as

*Given p and q integers, IF p ≠ q and p is prime and divides q,
THEN q is not prime.*

Now we can write it in the "*NOT B implies NOT A*" form.

   Given p and q are integers, IF NOT (q is not prime)
THEN
   NOT (p ≠ q and p is prime and p divides q)

Now observe that

   NOT (q is not prime)" = "q is prime

   NOT (A and B and C) = NOT(A) OR NOT(B) OR NOT(C)

Therefore,

   *NOT (p ≠ q and p is prime and p divides q) = NOT(p≠q) OR
NOT (p is prime) OR NOT (p divides q)*

   *Given p and q integers, IF q is prime, THEN (p=q OR p is not
prime OR p does not divide q)*

To prove this sentence, we can use the forward method starting
from the hypothesis, "*q is prime*".

It is sufficient to prove that at least one of "*p=q*", "*p is not prime*"
or "*p does not divide q*" must be true.

Now if $p = q$, we have proved the conclusion. Therefore, let us
assume that $p \neq q$.

If p is *NOT prime*, then we have proved the conclusion. Then let us assume p is prime.

Now if $p \neq q$ and p is prime, it cannot be "*p divides q*", because this would imply that q is NOT prime. Therefore, it must be that "*p does not divide q*".

In conclusion, we have proved that at least one of "*p = q*", "*p is not prime*", or "*p does not divide q*" will be true if the hypothesis is true.

Therefore, we have proved that the conclusion is true if the hypothesis is true.

9.14   Prove by contradiction, that there are at least two people on the planet who were born on the same second of the same hour of the same day of the same year in the twentieth century. (You can assume that there wee at least 4 billion people born into that century).

To prove "*A implies B*" by contradiction, we will put it in the form "*NOT B implies NOT A*" and then we will prove the statement to be true.

Let us put the statement in the "*A implies B*" form:

IF at least 4 billion people were born in the XXth century THEN there are at least two people on the planet who were born on the same second (of the hour of the day of the year) in the XXth century.

Now by the Contradiction Method, you can express it as:

NOT(there are at least two people on the planet who were born at the same second) IMPLIES NOT(at least 4 billion people were born in the XXth century)

Now we observe that

NOT (there are at least two people on the planet who were born at the same second) = "all people on the planet were born at DIFFERENT seconds in the XXth century"

And

*NOT(at least 4 billion people were born in the XXth century) =
"less than 4 billion people were born in the XXth century"*

And we want to prove the following:

*If "all people on the planet were born at DIFFERENT seconds in
the XXth century" THEN "less than 4 billion people were born in
the XXth century"*

This is simple using the forward method, because we start from
the hypothesis, and deduce from it the following intermediate
step:

Hypothesis:

*All people on the planet were born at DIFFERENT seconds in
the XXth century*

The hypothesis implies that the number of people born in the
planet during the XXth century is LESS THAN the number of
seconds in the XXth century.

Now as the number of seconds in a century is less than
100years*(366 days/year)*(24 hours/day)*(3600 seconds/hour
= 3.162.240.000 seconds, we have that:

The number of people born in the planet during the XXth
century is LESS THAN 3.162.240.000,

So, in the XXth century, there were less than 3.2 billion people
born. Then in the XXth century, there were also less than 4
billion people born, as we wanted to prove.

9.16 Prove, by contradiction, that if n is a positive integer such that $n^3 - n - 6 = 0$, then for every positive integer m with $m \neq n$, $m^3 - m - 6 \neq 0$.

To prove this by contradiction, we have to put "*A implies B*" under the form "*NOT B implies NOT A*". We start from this statement:

Given a positive integer n, IF ($n^3 - n - 6 = 0$) THEN for every positive integer m with $m \neq n$, $m^3 - m - 6 \neq 0$

Now let us translate it to "*NOT B implies NOT A*" notation, for example:

Given a positive integer n, IF NOT(for every positive integer m with $m \neq n$, $m^3 - m - 6 \neq 0$) THEN NOT ($n^3 - n - 6 = 0$)

Now "*for every positive*" is the same thing as "*FOR ALL positive*", and we have the rule that says:

NOT (FOR ALL x property something happens) is the same thing as "*THERE IS an x with such property such that something DOES NOT happen.*"

Therefore, *NOT(FOR ALL positive integer m with $m \neq n$, $m^3 - m - 6 \neq 0$)* is the same as:

THERE IS a positive integer m with $m \neq n$ such that NOT ($m^3 - m - 6 \neq 0$)

Which is the same as

THERE IS a positive integer m with $m \neq n$ such that $m^3 - m - 6 = 0$

Now we want to prove the following:

Given a positive integer n, IF THERE IS a positive integer m with $m \neq n$ such that $m^3 - m - 6 = 0$, THEN $n^3 - n - 6 \neq 0$

Note that the previous sentence is equivalent to this one:

Given two positive integers m and n, IF $m^3 - m - 6 = 0$ THEN $n^3 - n - 6 \neq 0$

Therefore, what we must prove is that THERE'S NO MORE THAN ONE ROOT for $m^3 - m - 6$ that is POSITIVE and an INTEGER.

To prove it, we compute the roots of $m^3 - m - 6$. One root is 2: $2^3 - 2 - 6 = 0$

Applying Ruffini to $m^3 - m - 6$ we obtain:

1	0	-1	-6
2	2	4	6
1	2	3	0

So $m^3 - m - 6 = (m - 2)(m^2 + 2m + 3)$

Therefore, if m is a root of $m^3 - m - 6$, it must be a root of $(m - 2)$ or a root of $(m^2 + 2m + 3)$, but we observe that both m and n are POSITIVE. It is **impossible** that they are roots of $m^2 + 2m + 3$, because all the terms should be positive.

Then m must be a root of $(x - 2)$, implying that $m = 2$. Therefore, as n is forced to be different than m (by hypothesis), it is forced to have a value that is not equal to 0, and so we have completed the proof.

9.19   Identify the contraction in the following proof.

Proposition. If a and b are integers with a ≠ 0 and the number of rational roots of $ax^4 + bx^2 + a$ is odd, then b is even.

To prove it by contradiction, we will try to prove the following:

Given a and b are integers with a ≠ 0, NOT(b is even) IMPLIES NOT(the number of rational roots of $ax^4 + bx^2 + a$ is odd).

Given a and b are integers with $a ≠ 0$: b is ODD IMPLIES the number of rational roots of $ax^4 + bx^2 + a$ is EVEN.

Now let us look how it to prove it.   Let us try to find the contradiction.

    (proof) *Assume, to the contrary, that b is odd.*

We assume that b is odd as this is our new hypothesis (remember that new hypothesis = NOT (original conclusion))

Then, by the proposition in the previous exercise, ± 1 are not roots of $ax^4 + bx^2 + a$.

We can deduce the following:

Suppose that 1 is the root: then we have: $a.1^4 + b.1^2 + a = 0$, and so,
$a + b + a = 0$, and so: $b = -2a$, and so b is even, which is in contradiction with our hypothesis that is: b is odd.

Therefore, it is impossible that 1 is the root of $ax^4 + bx^2 + a$. In the same way, it is impossible that (-1) is root of this polynomial, because $a(-1)^4 + b(-1)^2 + a = a.1^4 + b.1^2 + a$. For example, (-1) is root if and only if 1 is root, and we have already proved that 1 is not a root. Therefore, we have proved the result stated by the proof: +1 and -1 CANNOT be roots of $ax^4 + bx^2 + a$)

Now consider a rational root p/q of $ax^4 + bx^2 + a$

Here we start considering any rational root of $ax^4 + bx^2 + a$, and prove that there is always ANOTHER rational root, which is the different, for the polynomial. In this manner, we will prove that **the number of rational roots is even.**

Note that $p \neq 0$ for not $a = 0$.

(This observation is very simple: if $p = 0$, then $p/q = 0$, then 0 is a root of the polynomial, implying that a must be 0, but we know from the problem that $a \neq 0$, so p *cannot be 0*)

It is easy to verify that q/p is also a rational root of
$$ax^4 + bx^2 + a.$$

(As $p \neq 0$, we can define q/p. Now it is easy to observe that it is also a rational root of the polynomial by doing the following calculations:

We know that p/q is a rational root, then so we know that

$$a\,(p/q)^4 + b\,(p/q)^2 + a = 0.$$

Now as p/q is $\neq 0$, we can multiply at both sides of the equation by $(q/p)^4$ to get a new valid equation.

$$a + b\,(q/p)^2 + a\,(q/p)^4 = 0.$$

We observe that the equations implies that (q/p) is also a root of the polynomial $ax^4 + bx + a$, which is what it was stated above.

Thus, rational roots come in pairs.

That is true since we have proved that if (p/q) is a rational root of $ax^4 + bx^2 + a$, THEN (q/p) is also a rational root of $ax^4 + bx^2 + a$

At this moment our proof for the contradiction is complete, and we have proved that "*NOT b implies NOT a*".

Because the number of rational roots of $ax^4 + bx^2 + a$ is odd, it must be that one of these roots is repeated, so $p/q = q/p$.

Here contradiction was used to prove "*A implies B*".
We know that "*NOT B implies NOT A*".

Therefore, if we start from A, we then know that the roots p/q and q/p must be equal.

But then $p = \pm\, q$, which cannot happen, and so the proofs is complete.

And so p/q and q/p must be +1 and -1. And as we have proved previously, this cannot be the roots of the polynomial. Here is the contradiction, since if we start from not b, we obtain something that is in contradiction with a. Therefore, b must be true if a is true.

Chapter 10
The Contrapositive Method

10.2   Bob said, "If I study hard, then I will get at least a B in this course." After the course was over, Mary said, "You got an A in the course, so therefore you studied hard. "Was Mary's statement correct?

Mary's statement is incorrect.   Because Bob said: *"If I study hard, then I will get at least B in this course."*   He said nothing about what will happen if he does NOT study hard.

For example, based on Bob's statement, it is POSSIBLE that even if he does NOT study hard, he will get an A or a B in this course. Bob only said: *"IF I STUDY HARD ...,"*   But he said nothing about what happens if he does not study hard.

Therefore, when Bob got an A in the course, can Mary conclude that he studied hard? NO!

Mary is NOT correct since she assumed that obtaining an A only could happen if Bob studied hard.

10.4   Suppose that a, b, and c are real numbers with $a > 0$ and that you want to prove, by the contrapositive method, that "If there is no real number x with $ax^2 + bx + c = 0$, then it is not true that for all real numbers x, $ax^2 + bx + c < 0$." Which of the Construction, Choose or Specialization Techniques would you use to prove this?

The problem is in the form *"IF A then B"*.   To prove by the contrapositive method, first try to rewrite the statement:

If there is no real number x, with $ax^2 + bx + c = 0$, then it is not true that for all real numbers x, $ax^2 + bx + c < 0$, **then** *ASSUME NOT B and A and try to prove NOT A.*

To prove the following:

IF there is no real number x with $ax^2 + bx + c = 0$
THEN it is not true that for all real numbers x, $ax^2 + bx + c < 0$

A = "there is no real number x with $ax^2 + bx + c = 0$"
B = "it is not true that for all real numbers x, $ax^2 + bx + c < 0$"

Using the contrapositive method

> **ASSUME** that: A and NOT B are true
> Then **PROVE** that: NOT A is true

Therefore, there is a contradiction, because *A and NOT B* cannot be true at the same time, which means that NOT B must be false, and so *B* is true if *A* is true.

Now assume the following:

> **A** = "there is no real number x with $ax^2 + bx + c = 0$"
> **NOT B** = NOT("it is not true that for all real numbers x, $ax^2 + bx + c < 0$)".

Then try to conclude the following:

> **NOT A** = NOT(there is no real number x with $ax^2 + bx + c = 0$)".

Rewriting *NOT B*:

> **NOT B** = "for all real numbers x, $ax^2 + bx + c < 0$".

And examine the following:

> **NOT A** = THERE IS a real number x with $ax^2 + bx + c = 0$

Now use the construction method to build an *x* such that $ax^2 + bx + c = 0$.

1: Use the fact that $a > 0$ to conclude that there must be an x_1 such that $ax_1^2 + bx_1 + c > 0$.

2: Use the fact that *NOT B* is assumed to be true, to prove that there must be an x_2 such that $ax_2{}^2 + bx_2 + c < 0$.

3: Use the fact that $ax^2 + bx + c$ is a *continuous* function and the theorem of the intermediate value to conclude that *there must be an x between x_1 and x_2 such that $ax^2 + bx + c = 0$.*

Therefore, the desired x to satisfy *NOT A* is constructed proving that *NOT A* is true.

Thus, by the contrapositive method, "*NOT B*" cannot be true if "*A*" is true. Therefore, B must be true if A is true.

10.6 Suppose that f is a function of one variable, S is a set of real numbers, and that the contrapositive method is being used to prove the proposition, "If no element x Element of S satisfies the property that f(x) = 0, then f is not bounded above."

Which of the following is the correct key question? What is wrong with the other choices?

a. How can I show that a function is not bounded above?

The *Key Question, "How can I show"* is the conclusion wanted. In case (a), we need to conclude *NOT A* - *"there is an element x of S that satisfies f(x) = 0"*. Thus, the question asked is not useful. Therefore, (a) is **NOT** correct.

b. How can I show that there is an element x Element of S such that f(x) = 0?

The *Key Question "How can I show"* is the conclusion wanted. In case (b), we need to conclude *NOT A* - *"there is an element x of S that satisfies f(x) = 0"*. Thus, in this case, the *Key Question* is **CORRECT**.

137

c.   How can I show that a function is bounded above?

The *Key Question* "*How can I show*" is the conclusion wanted. In case (c), need to conclude *NOT A* - "*there is an element x of S that satisfies f(x) = 0*". Thus, the question asked is not useful. Therefore, (a) is **NOT** correct.

d.   How can I show that there is a point in a set where the value of a function is 0?

The *Key Question* "*How can I show*" is the conclusion wanted. In case (d), we need to conclude *NOT A* - "*there is an element x of S that satisfies f(x) = 0*". Thus, the question asked here is useful, because this key question is an intermediate step:

First, from the question we know that there is a point in a set where the value of a function is 0, and in a further step, this fact is used to prove that there is an x in S such that f(x) = 0. Thus, in this case, the *key question* is correct.

10.8   Suppose that a and b are positive real numbers. Prove, by the contrapositive method, that if a ≠ b, then
$(a+b)/2 > (ab)^{.5}$

Using the contrapositive method, prove that "*A implies B*":

Assume A and NOT B are true

This can be proved based on the fact that *NOT A is true (contradiction with A)*, thus, the hypothesis (*NOT B* was true) is wrong, so *B* must be true if *A* was true.

To start, we want to prove that given *a* and *b* which are positive real numbers,

$$if\ a \neq b\ then\ (a+b)/2 > (ab)^{1/2}$$

Now assume the following:

$$A = "a \neq b"$$
$$NOT\ B = NOT((a+b)/2 > (ab)^{1/2}) = (a+b)/2 <= (ab)^{1/2}$$

Then conclude the following:

$$NOT\ A = NOT\ (a \neq b) = \text{“}a = b\text{”}$$

Start from "$(a+b)/2 \leq (ab)^{1/2}$", and prove: $a = b$

$$If\ (a+b)/2 \leq (ab)^{1/2},\ then\ (a+b) \leq 2(ab)^{1/2}$$

Since a and b are positive, and the function $f(x) = x^2$ is strictly increasing for $x > 0$, then

$$(a+b)^2 \leq (2(ab)^{1/2})^2$$

Therefore, $(a^2 + 2ab + b^2) \leq 4ab$, and $(a^2 - 2ab + b^2) \leq 0$, and $(a - b)^2 \leq 0$.

And as $x^2 \geq 0$ for all x real, and only $0^2 = 0$, it must be that $(a - b)^2 = 0$. Therefore, $(a - b) = 0$, and $a = b$.

So it is proved that *NOT A* is true. Thus, by the *contrapositive method*, the conclusion is that "*A implies B*". If we assumed that *A and NOT B*, we get *NOT A*, and this is in contradiction with the assumption.

10.10   Use the approach in the proof of Proposition 15 to prove that the function $f(x) = x^3$ is one-to-one.

Proposition 15: *If m and b are real numbers with m $\neq 0$, then the function f(x) = mx+b is one-to-one.*

Proof of Proposition 15: *Let x and y be real numbers for which mx+b = my + b. It is shown that x=y. But this follows by subtracting b from both sides and then dividing by m, noting m$\neq 0$.*

We let x and y be real numbers for which $f(x) = f(y)$. We must show that $x = y$ to conclude that f is one-to-one.

Since it is known that $x^3 = y^3$, then $x^3 - y^3 = 0$, and $(x - y)(x^2 + xy + y^2) = 0$, because $(x - y)(x^2 + xy + y^2) = x^3 - y^3$.

Therefore, if x and y are real numbers such that $f(x) = f(y)$, then
$$(x - y)(x^2 + xy + y^2) = 0$$

Since $x^2 + xy + y^2$ is always > 0, because this polynomial has no roots, as can be verified by applying the formula to solve polynomials of degree 2. Therefore, to satisfy $(x - y)(x^2 + xy + y^2) = 0$, it must be that $x - y = 0$.

Therefore, if x and y are real numbers such that $f(x) = f(y)$, then $x - y = 0$, then $x = y$, and f is a one-to-one function.

10.12 Suppose that v is an upper bound for a set S of real numbers. Prove, by the contrapositive method, that if it is not true that there is a real number e > 0 such that for every element x element of S, x ≤ v - e, then there is no real number u < v such that u is an upper bound of S.

By the *contrapositive* method, we can prove that "*A implies B*" by doing the following:

Assume A and NOT B are true

We can prove based on the following:

That *NOT A* is true (contradiction with *A*, so our hypothesis *NOT B* was true, is wrong. Thus, *B* must be true if *A* was true)

To start, we want to prove that

Given a set S of real numbers, and v as an upper bound for S IF it is not true that there's a real number e >0 such that for every element x of S, x≤ v– e THEN there is no real number u < v such that u is an upper bound of S.

Let us assume the following:

A= "it is not true that there's a real number e>0 such that for every element x of S, x≤v– e".

NOT B = NOT (there is no real number u < v such that u is an upper bound of S).

Then we must conclude the following:

NOT A = NOT (it is not true that there is a real number e > 0 such that for every element x of S, x ≤ v - e) = "there is a real number e > 0 such that for every element x of S, x ≤ v − e".

Therefore, starting from "NOT (there is no real number u < v such that u is an upper bound of S)", must prove that: "there is a real number e > 0 such that for every element x of S, x ≤ v − e".

To begin, start from what is known

NOT(there is no real number u < v such that u is an upper bound of S) : There is a real number u < v such that u is an upper bound of S.

We want to reach a *THERE IS* statement. Therefore, use the *construction* method to prove it:

Consider e = (v - u)/2

Observe that e > 0 because u < v by hypothesis, and that if x is in S, then x ≤ u (because by hypothesis, u is an upper bound for S), and so x ≤ u ≤v − e; the first inequality, because u is an upper bound for S; and the second inequality is that v − e = v − (v − u)/2 = (v+u)/2, and u ≤ (v+u)/2 because v ≥ u.

This is the wanted conclusion: x ≤ v − e

Therefore, starting from *A and NOT B*, the conclusion is reached that *NOT A* must be *false*.

Then, using the contrapositive method, the conclusion is reached that if *A is true*, then *B must be true.* Therefore, *A implies B*.

10.14 Prove, by the contrapositive method, that if u is a least upper bound for a set S of real numbers, then for all real numbers e > 0, there is an element x Element of S such that x > u − e.

Using the *contrapositive method*, we can prove that "*A implies B*" by doing the following:

First assume that *A and NOT B* are true.

The proof is based on the following:

NOT A is true (contradiction with A, so our hypothesis NOT B was true is wrong, therefore, B must be true if A was true).

To start, we want to prove that

Given a set S of real numbers
If u is a least upper bound for S
Then for all real numbers e > 0, there is an element x of S such that x > u − e.

Let us assume the following:

A = "u is a least upper bound for S"
NOT B = NOT (for all real numbers e > 0, there is an element x of S such that x > u − e)

Therefore, we must conclude that

NOT A = NOT (u is a least upper bound for S) = "u is not a least upper bound of S"

So starting from "*NOT(for all real numbers e > 0, there is an element x of S such that x > u - e)*", we must prove that "*u is not a least upper bound for S*"

To begin, start from what is known.

NOT(for all real numbers e > 0, there is an element x of S such that x > u − e).
THERE IS a real number e>0 such that NOT (there is an element x of S such that x > u − e).

There is a real number e > 0 such that FOR ALL x in S,
NOT(x > u − e).

There is a real number e > 0 such that for all x in S, x ≤ u − e

This is our hypothesis. Now by the forward method, we try to prove the desired conclusion "*u is not a least upper bound of S*".

Now our hypothesis: "*THERE IS a real number e > 0...*"

So considering such an *e, an e > 0* such that for all *x in S,*
x ≤ u − e.

Therefore, *u − e* is an upper bound for S (because this is the definition of upper bound: U is an upper bound of S if $x \le U$ for all x in S, so u − e is an upper bound for S).

And since *u − e < u,* then *u − e* is an upper bound for *S,* and *u − e< u.* Now putting all this together:

u is NOT the least upper bound of S (because the least
upperbound is the least of all upper bounds, and above it is
proved that there's another upper bound, u − e, that is less than
u)

Then the conclusion is reached that *u* is not the least upper bound of S. Therefore, starting from *A and NOT B*, it is proved that *NOT A* must be false. Thus, by using the *contrapositive* method, we have proved that if *A* is true, then *B* must be true. Therefore, *A* implies *B*.

143

10.16 Write an analysis of proof that corresponds to the condensed proof given below. Indicate which techniques are used and how they are applied. Fill in the details of any missing steps where appropriate.

Definition: *A set S of real numbers is bounded if and only if there is a real number M > 0 such that, for all elements x element of S, $|x| < M$.*

The definition states that a set S of real numbers is BOUNDED, if there is a way to put all its elements between $-M$ and M, such that for all x in S, $-M < x < M$ (equivalent to the definition $|x| < M$).

Proposition: *Suppose that S and T are sets of real numbers with S subset of T. If S is not bounded, then T is not bounded.*

Proof: *Suppose that T is bounded. Hence, there is a real number $M' > 0$ such that, for all x element of T, $|x| < M'$. It is shown that S is bounded. To that end, let x' Element of S. Because S is a subset of T, it follows that X' element of T. But then $|x'| < M'$ and so S is bounded, thus completing the proof.*

The proposition states that if *S is a subset of T*, and *S is NOT bounded*. Then *T* cannot be BOUNDED, which is true.

Now from the **proof**: *Suppose that T is bounded.*

Start by using the contrapositive method, and

1: Assume that A and NOT B are true

2: Based on this, try to conclude that NOT A must also be true (and this conclusion is in contradiction with the assumption – so B must be true if A was true).

Now from the *Proposition*: *"if S is not bounded, then T is not bounded".*

> *A = "S is not bounded"*
> *B = "T is not bounded"*

Assuming **NOT** (B), then **NOT** ("T is not bounded") = "T is bounded" *by the contrapositive method.*

This is exactly what the given proof states: "*suppose that T is bounded.*"

(proof) Hence, there is a real number M'>0 such that, for all x element of T, |x| < M'...

Now, we go forward one step, and apply the *definition* of being bounded: *a set is bounded if there exists an M > 0 so that for all x in the set, |x| < M.*

In our case, the set is T and M' the bounding value. Then, since *T is bounded*, apply the definition of being bounded and conclude:

There is a real number M' > 0 such that, for all x of T, |x| < M'

(proof) It is shown that S is bounded...

The goal is to use the contrapositive method and to *prove NOT(A) = NOT (S is not bounded) = "S is bounded".* If it is proved, then it is done.

Now the following line proves that *S* is bounded:

(proof) To that end, let x' Element of S...

Now, in trying to prove that S is bounded, we must prove that S satisfies the definition of being bounded: **there is an M > 0 such that, for all x in S, |x| < M.**

Now to prove a "*THERE IS*" statement, we use the *construction method.*

The goal is to construct (find) an M such that, for all x in S, |x| < M. Now, to prove a "*for all*" statement, use the *choose method.* Consider an arbitrary x from S, and try to prove that it has the desired property.

Therefore, as is written in the proof: *consider an x' element of S...*

(proof) Because S is a subset of T, it follows that X' element of T...

For this x' in S that was previously considered, we can arrive at the conclusion: ***x' is also in T, because S is included in T.***

(proof) But then $|x'| < M'$ ...

Now since it is known that x is in T, $|x| < M'$, then for this x' in S (that is also in T), it will be true that: $|x|' < M'$

(proof) and so S is bounded ...

Yes! because since we previously proved that: ***for any x' in S, $|x|' < M'$.***

Therefore, it is proved that *for all x' in S, $|x'| < M'$*, then S is bounded (by *definition* of being bounded).

(proof) thus completing the proof.

Thus, the proof is complete. We have proved that S *is bounded*, and Proved *NOT(A)*, the desired conclusion.

We started by **ASSUMING** *A and NOT(B)*, and concluded *NOT(A)*, which is in contradiction with *A*. Which means that *NOT(B)* and *A* are incompatible, that *A* implies *B*, as we wanted to prove.

Chapter 11
The Induction Method

_11.2_   Prove that if a and b are integers, with $a \neq 0$, such that a/b, then there is a unique integer k such that $b = ka$.

We will try to proving this using the forward method. We will start from the hypothesis, and go to the conclusion (step by step).

Let us start with the hypothesis:

a and b are integers with a $\neq$ 0 such that a / b

The definition of division, "a / b", means that

there is an integer k such that a*k = b

Therefore, by the definition of "_division_", a/b implies that there is an integer k such that ak = b.

To complete the proof, we just have to prove that k is unique.

Let us assume, by contradiction, that there is another k', $k \neq k'$, such that $ak' = b$. Thus, we have that $ak = b$ and $ak' = b$.

Then $ak - ak' = b - b = 0$, and $(k - k') = 0$. And as $a \neq 0$, it must be that $k - k' = 0$, and k = k'. This is in contradiction with our initial assumption, that k was different from k'. Therefore, the assumption is false, and it must be that $k = k'$, and k is unique.

Thus we have proved the conclusion that there is a unique integer k such that $ak = b$.

11.4   Prove, by the indirect uniqueness method, that there is a unique integer n for which $2n^2 - 3n - 2 = 0$.

The *indirect uniqueness* method is the following:

Suppose by contradiction there are two different integers n and n' such that the property holds, and then prove that n = n' (that is, the integer n is unique).

Let us assume then that there are two different integers n and n' such that:

$$2n^2 - 3n - 2 = 0$$
$$2n'^2 - 3n' - 2 = 0.$$

If we subtract the second equation from the first, we get the following:

$$2n^2 - 3n - 2 - (2n'^2 - 3n' - 2) = 0$$
$$2(n^2 - n'^2) - 3(n - n') - 2 + 2 = 0$$

$$2(n^2 - n'^2) - 3(n - n') = 0$$
$$2(n - n')(n + n') - 3(n - n') = 0$$
$$(because\ n^2 - n'^2 = (n - n')(n + n'))$$
$$(n - n')[2(n + n') - 3] = 0.$$

We can put $(n - n')$ as a common factor, and this implies that:

$$2(n + n') - 3 = 0$$
$$(n + n') = 3/2.$$

Therefore, both n and n' are integers, and their addition is not an integer. This is impossible, and so our initial assumption was wrong that there are no two integers that verify the equation $2n^2 - 3n - 2 = 0$.

So there is a 0 or 1 integer that verifies this equation. Now we observe that n = 2 verifies the equation because

$$2 * 2^2 - 3*2 - 2 = 2*4 - 6 - 2 = 8 - 8 = 0.$$

And so we have found that $n = 2$ is a unique integer that verifies the given equation.

11.7 Upon learning about the method of induction, a student said, "I do not understand something. After showing that the statement is true for n=1, you want me to assume that P(n) is true and to show that P(n+1) is true. How can I assume that P(n) is true - after all, aren't we trying to *show* that P(n) is true? Answer this question.

The student did not understand the true meaning of the induction method, which is based on these two assertions:

1) P(1) is true.
2) If P(n), then P(n+1) is true.

Note that the statement that is used in the "induction step", is an "IF... THEN..." statement.

This kind of statement asserts that *"if we assume that a condition is verified, then a conclusion can be asserted."*

It is not true what the student said that, *"How can I assume that P(n) is true – when I'm trying to prove that P(n) is true"*. The induction step does not assume that *P(n)* is true.

The student is also confused about the following:

He assumes P(n) is true FOR ALL n.

The induction step just said that if *P(n)* then *P(n+1)*, where *n* is a certain n. That is if *P(2)* then *P(3)*, and if *P(3)* then *P(4)*.

How the "*n*" is used in both expressions of *P(n)* is different. In the first case, it is used in the sense of "*any n you want*". In the second case, it is used in the sense of "*a given n*".

11.9   Suppose that A(n) and B(n) are statements that depend on a positive integer n. Explain how you would use induction to prove that for every integer n ≥ 1, if A(n) is true, then B(n) is true. For Step 2 of the induction, indicate what statements you would assume are true and what statements you would have to show are true.

Remember that in the induction method, to prove that P(n) is true for n ≥ 1, we have to do the following steps:

BASE STEP: Prove that P(1) is true.

INDUCTION STEP: Prove that "IF P(n) THEN P(n+1)" is valid for all n ≥ 1.

In our case, we want to prove that if *A(n)* is true, then *B(n)* is true for all n.

Therefore, we can say that P(n) = "if A(n) is true, then B(n) is true". And so to apply the induction method to this case, we try to prove by following these steps:

BASE STEP: Prove that "IF A(1) is true, THEN B(1) is true"

INDUCTION STEP: Prove that "if (if A(n) is true, then B(n) is true), and then (if A(n+1) is true, then B(n+1) is true)", or, expressed in a more friendly language, the induction step will be:

Prove that:
> *"if A(n) is true, then B(n) is true"*
> *implies that*
> *"if A(n+1) is true, then B(n+1) is true".*

In the induction step, we must assume the following to be true:

> *"If A(n) is true, then B(n) is true"*
> *and we must show as true:*
> *"If A(n+1) is true, then B(n+1) is true".*

11.11 Prove, by induction, that for every integer $k \geq 1$,
$$1 + 2 + 2^2 + \ldots + 2^{k-1} = 2^k - 1.$$

1) **BASE STEP**:

We must prove that $1 + 2 + \ldots + 2^{(1-1)} = 2^1 - 1$, that is, $1 = 2^1 - 1$, and that $1 = 2 - 1$, is TRUE.

2) **INDUCTION STEP**:

We must prove that, if $1 + 2 + 2^2 + \ldots + 2^{(k-1)} = 2^k - 1$ is true, then $1 + 2 + 2^2 + \ldots + 2^{(k-1)} + 2^k = 2^{(k+1)} - 1$ is also true.

So in our induction hypothesis we assume that

$$1 + 2 + 2^2 + \ldots + 2^{(k-1)} = 2^k - 1$$

And based on this, we want to prove that

$$1 + 2 + 2^2 + \ldots + 2^{(k-1)} + 2^k = 2^{k+1} - 1.$$

Then, going forward from the induction hypothesis, we assume as true:

$$1 + 2 + 2^2 + \ldots + 2^{k-1} = 2^k - 1.$$

If we add 2^k to both sides, we get the following:

$$1 + 2 + 2^2 + \ldots + 2^{k-1} + 2^k = 2^k - 1 + 2^k$$
$$1 + 2 + 2^2 + \ldots + 2^{k-1} + 2^k = 2 * 2^k - 1.$$

And using the fact that $2 * 2^k = 2^{k+1}$, then

$$1 + 2 + 2^2 + \ldots + 2^{k-1} + 2^k = 2^{k+1} - 1$$

Therefore, we have proved the induction step, and we have completed the proof.

11.13   Prove, by induction, that for every integer
          $n \geq 1$, $1/n! \leq 1/2^{n-1}$

1) **BASE STEP**:

We must prove that $1/1! \leq 1/2^{(1-1)}$, that is, that $1 = 1/2^0$, and $1 = 1$, TRUE.

2) **INDUCTION STEP**:

We must prove that

> *If $1/n! \leq 1/2^{n-1}$ is true*
> *then $1/(n+1)! \leq 1/2^n$ is also true*

We assume that $1/n! \leq 1/2^{n-1}$ (*induction hypothesis*)

And based on this, we want to prove that $1/(n+1)! \leq 1/2^n$

Let us try to do it by going forward from the induction hypothesis. We assume as true $1/n! \leq 1/2^{n-1}$.

Now we observe that $1/(n+1) \leq 1/2$. Therefore, if we multiply the left side by $1/(n+1)$, and the right side by $1/2$, the "$\leq$" symbol will be maintained in our original equation, and so we obtain the following:

$$(1/n!)(1/(n+1)) \leq (1/2^{n-1}) * 1/2$$

Now we observe that the left side is $1/(n+1)!$, and the right side is $1/2^n$, and so we have proved that $1/(n+1)! \leq 1/2^n$.

11.15   Prove, by induction, that if x > 1 is a given real number, then for every integer n ≥ 2, $(1+x)^n > 1 + nx$

1) **BASE STEP**:

We must prove that the statement is true for $n = 2$ (as we want to prove the property for $n \geq 2$). So we must prove that

$$(1+x)^2 > 1 + 2x.$$

That is

$$1 + 2x + x^2 > 1 + 2x$$
$$x^2 > 0, \text{ TRUE because } x > 1.$$

2) **INDUCTION STEP**:

We must prove that

IF	$(1+x)^n > 1 + nx$	is true
THEN	$(1+x)^{n+1} > 1 + (n+1)x$	is also true

So we assume that
$$(1+x)^n > 1 + nx$$
$$(induction\ hypothesis).$$

And based on it we want to prove that

$$(1+x)^{n+1} > 1 + (n+1)x$$

Then going forward from the induction hypothesis, we assume as true
$$(1+x)^n > 1 + nx.$$

Now if we multiply by both sides by (1+x), we get the following:

$$(1+x)(1+x)^n > (1+x)(1 + nx).$$

Now as $nx^2 > 0$, we have that $\quad (1 + x + nx + nx^2) > (1 + x + nx)$

So we have
$$(1+x)^{n+1} > 1 + x + nx + nx^2 > 1 + x + nx$$
$$(1+x)^{n+1} > 1 + (n+1)x.$$

So, we have proved the induction step. And so we have completed the proof.

11.17 A machine is filled with an odd number of chocolate candies and an odd number of caramel candies. For 25 cents, the machine dispenses two candies. Prove that, before being empty, the machine will dispense at least one pair that consists of one chocolate candy and one caramel candy.

Let N and M be the number of chocolate and caramel candies that are initially in the machine, and we know that both N and M are odd.

Suppose then, by contradiction, that there is a way to provide pairs of candies that are always "only chocolate" or "only caramel". After the first pair is provided, the number of candies in the machine will be N' and M', where:

$$N' = N - 2$$
$$M' = M$$

$$OR$$

$$N' = N$$
$$M' = M - 2$$

In both cases, N' and M' are still odd numbers.

Now, we continue with this process. In each step, we assume that there is a pair of only caramel or only chocolate. In each step, N and M continue to be odd numbers. And eventually, N or M will reach the value $N = 1$ or $M = 1$ (because they are going down by odd numbers).

Then the next pair will have 1 caramel candy and 1 chocolate candy, which is in contradiction with our initial assumption.

Therefore, there is no way to empty the machine without dispensing at least one pair of only one chocolate and one caramel candy.

11.19 Prove that in a line of at least two people, if the first person is a woman and the last person is a man, then somewhere in the line there is a man standing immediately behind a woman.

We will do the proof by the induction method on the size n, for n ≥ 2.

1) **BASE STEP**:

We must prove for $n = 2$ if the line has two persons. The first person is a woman and the last person is a man. Then somewhere in the line there is a man standing immediately behind a woman. This statement is obviously TRUE, as the first is a woman, the last is a man, and there are just two in the line.

2) **INDUCTION STEP**:

We must prove that if the assertion is true for n, then it is true for $n+1$. We must prove the following:

IF it is true that in a line of n persons, starting with a woman and ending with a man, there is always a man immediately behind a woman.

THEN it is true that in a line of n+1 persons, starting with a woman and ending with a man, there is always a man immediately behind a woman.

We will do the proof of the induction step by the forward method. We will start from the assumption and prove the conclusion.

Let us start from the following statement:

In a line of n persons, starting with a woman and ending with a man, there is always a man immediately behind a woman.

Now, we want to prove that the assertion is also true for a line of (n+1) persons.

155

To do it, let us consider the line of (n+1) persons, and then consider the following:

1) The first woman
2) The line of n persons that come after the first woman

In step (2), two things may happen:

a) Line (2) starts with a MAN. In this case, the conclusion is proved, because there is a man immediately behind a woman.

b) Line (2) starts with a WOMAN. In this case, we have that line (2) satisfies the conditions of the induction hypothesis. Therefore, there is by the induction hypothesis, a man immediately after a woman in line (2), and again we have proved the conclusion is true.

Therefore, in all cases, we have proved the conclusion of the induction step that *"there is a man immediately after a woman at some place in the line of n+1 persons."*

11.21 Describe a modified induction procedure for proving that for every positive odd integer n, P(n) is true.

The normal induction procedure is the following:

1) **BASE STEP**:   *Prove P(1)*

2) **INDUCTION STEP**:   *Prove that P(n) implies P(n+1)*

Therefore, the base step proves P(1), and the induction step proves P(2) (because P(1) was true), P(3) (because P(2) was true), P(4), P(5), P(6), and so on. In our case, we want to prove P(1), and then prove P(3), P(5), P(7), etc.

An alternative induction procedure – where the results are restricted to odd numbers - is the following:

1) **BASE STEP**:   Prove P(1)

2) **INDUCTION STEP**:   Prove that P(n) implies P(n+2)

The base step proves P(1), and the induction step proves P(3), P(5), P(7), etc.

An even better general solution (in the sense of requiring less to be proved) is:

1) **BASE STEP**:   Prove P(1)

2) **INDUCTION STEP**:   Prove that, if n is odd,
                                                P(n) implies P(n+2)

11.24 In the following condensed proof, explain how the author relates $P(n+1)$ to $P(n)$, and where in the proof the induction hypothesis is used.

Proposition: *For every integer $n \geq 2$, if x_1, x_2, ... x_n are real numbers > 0 and < 1, then $(1-x_1)(1-x_2)...(1-x_n) > 1 - x_1 - x_2 - ... x_n$.*

This is the proposition to be proved. Note that it is to be proved for $n \geq 2$, and so when using the induction method, we will have to do the following:

1) **BASE STEP**:

When $n = 2$, prove that $x1$, $x2$ are real numbers > 0 and < 1, then $(1 - x1)(1 - x2) > 1 - x1 - x2$.

2) **INDUCTION STEP**:

Prove that if the equation is true for n, then it is true also for $n+1$.

Now we will look at how we find each of these two steps in the proof given with the problem.

Proof: *When $n = 2$, the statement is true because you have that $(1-x_1)(1-x_2) = 1 - x_1 - x_2 + x_1 x_2 > 1 - x_1 - x_2$. Assume the statement is true for n. Then,*

$$(1-x_1)(1-x_2)...(1-x_{n+1})$$
$$= \quad [(1-x_1)(1-x_2)...(1-x_n)](1-x_{n+1})$$
$$> \quad (1 - x_1 - ... - x_n)(1 - x_{n+1})$$
$$= \quad 1 - x_1 - ... x_n - (1 - x_1 - x_2 - ... -x_n)(x_{n+1})$$
$$= \quad 1 - x_1 - ... x_n - x_{n+1} + (x_1 + x_2 + ... + x_n)(x_{n+1})$$
$$> \quad 1 - x_1 - ... -x_n - x_{n+1}$$

The proof is now complete.

The proof starts with the case $n = 2$, as we guessed initially, because this is the base step. Therefore, this $n = 2$ case is the base step for the induction proof.

In the induction step, we must prove *that*

$$(1 - x_1)(1 - x_2) > 1 - x_1 - x_2.$$

Which is true, because

$$(1 - x_1)(1 - x_2) = 1 - x_1 - x_2 + x_1 x_2.$$

And as $x_1 x_2 > 0$

$$1 - x_1 - x_2 + x_1 x_2 > 1 - x_1 - x_2.$$

And so

$$(1 - x_1)(1 - x_2) > 1 - x_1 - x_2, \text{ as was stated.}$$

Let us assume the statement is true for n. Then the induction step starts here. We assume that the statement is true for n, and have to prove that it will be true for n+1.

Therefore, to begin proving that it is true for n+1, we want to prove that

$$(1 - x_1) \ldots (1 - x_{(n+1)}) > 1 - x_1 - x_2 - \ldots - x_n - x_{(n+1)}$$

It is true that

$$(1 - x_1) \ldots (1 - x_n) > 1 - x_1 - x_2 - \ldots - x_n$$

$$(1-x_1)\,(1-x_2) \ldots (1-x_{n+1}).$$

First, the left side of the conclusion is expressed. We want to prove that it is greater than $1 - x_1 - x_2 - \ldots - x_n - x_{(n+1)}$, which will be done in the lines that follow

$$= \qquad [(1-x_1)\,(1-x_2) \ldots (1-x_n)]\,(1-x_{n+1}).$$

Just done the n first terms (to make more apparent the left term of the equation in the induction hypothesis, that is
$$(1 - x_1) \ldots (1 - x_n) > 1 - x_1 - \ldots - x_n)$$

Therefore, as

$$(1 - x_1) \cdot \ldots \cdot (1 - x_n) > 1 - x_1 - \ldots - x_n.$$

It is the same as multiplying both sides by $(1 - x_{(n+1)})$.

$$= \qquad 1 - x_1 - \ldots x_n - (1 - x_1 - x_2 - \ldots - x_n)(x_{n+1})$$

$$= \qquad (1 - x_1 - \ldots - x_n) * 1 - (1 - x_1 - \ldots - x_n)(x\_n+1)$$

$$= \qquad 1 - x_1 - \ldots x_n - x_{n+1} + (x_1 + x_2 + \ldots + x_n)(x_{n+1})$$

The term "$- x_{n+1}$" comes from the first term of $(1 - x_1 - \ldots - x_n)$ (x_n+1). The rest is as before.

$$> \qquad 1 - x_1 - \ldots - x_n - x_{n+1}$$

Observe that $(x_1 + \ldots + x_n)(x\_n+1) > 0$, so the assertion is true, and the proof is now complete, because we have proved that

$$(1 - x_1) \cdot \ldots \cdot (1 - x\_n+1) > 1 - x_1 - \ldots - x\_n+1$$

11.27 What if anything is wrong with the following condensed proof?

Proposition: *For every integer n ≥ 2, if S₁, S₂, ..., Sₙ are convex sets of real numbers, then S₁ U S₂ U ... Sₙ is a convex set.*

This assertion is FALSE. Therefore, the proof given below by the problem must be wrong.

Proof: *To see that the statement is true for n=2, let x, y element of S₁ U S₂ and let t be a real number with 0 ≤ t ≤ 1.*

Because x,y element of S₁ U S₂ and because S₁ is convex, tx+ (1-t)y element of S₁. Likewise, because S₂ is convex, tx+ (1-t)y element of S₂. Thus tx+(1-t)y element of S₁ U S₂ and so the union of two convex sets is a convex set. Assume now that the statement is true for n and let S₁, S₂, ... , Sₙ₊₁ be convex sets of real numbers. You then have that:

$$S_1 \; U \;.... \; U \; S_{n+1} = (S_1 \; U \;... \; U \; S_n) \; U \; S_{n+1}$$

By the induction hypothesis, S = S₁ U ... U Sₙ is convex. Finally, it has already been shown that the union of two convex sets is convex and therefore S U S ₙ₊₁ is convex, completing the proof.

1) **BASE STEP**: (the case where n = 2)

We have to prove that *if S₁, S₂ are convex sets of real numbers, then S₁ U S₂ is a convex set.*

(NOTE: this base step is false since the proof above has an error.)

The proof provided tries to prove that if *S₁ and S₂ are convex sets, then S₁ U S₂ is also a convex set.* To do this, it uses the definition of convex set:

A set S is convex if and only if for all x, y in S, then tx + (1-t)y is in S, where t is between 0 and 1.

Because *x and y* element of *S₁ U S₂* and because S₁ is convex, tx+ (1-t)y is an element of S₁.

The following is the error in the proof:

It is not true that $tx + (1-t)y$ is an element of S_1, because we do not know for sure that x or y are in S_1. We only know that x and y are in $S_1 \cup S_2$, but this does not imply that they are in S_1. They may be in S_2. Thus, it is false that $tx + (1-t)y$ is in S_1. And therefore, the rest of the proof is wrong.

(proof) Likewise, S_2 is convex, $tx + (1-t)y$ element of S_2.

The same error is made here. It is NOT true *that $tx + (1-t)y$ is in S_2.*

(proof) Thus, $tx + (1-t)y$ element of $S_1 \cup S_2$, and so the union of two convex sets is a convex set.

This is also wrong. Because if $tx + (1-t)y$ is in S_1 and S_2, then it is in $S_1 \cup S_2$.

The base step failed. Therefore, the proof is wrong.

(proof) Assume now that the statement is true for n and let S_1, S_2, ... , S_{n+1} be convex sets of real numbers. You then have that
$$S_1 \cup .... \cup S_{n+1} = (S_1 \cup ... \cup S_n) \cup S_{n+1}$$

By the induction hypothesis, $S = S_1 \cup ... \cup S_n$ is convex. Finally, it has already been shown that the union of two convex sets is convex and therefore $S \cup S_{n+1}$ is convex, completing the proof.

Since we have found errors in the base step and in the induction step, the entire proof is wrong.

Chapter 12
Min/Max Methods

Proof by Cases: Suppose that the key words _either/or_ arise in the hypothesis of a proposition whose form is as "_D OR E implies B._"

According to the forward-backward method, to assume that _D OR E_ is true; you must conclude that _B is true._

The only question is whether you should assume that _D is true_ or whether you should assume that _E is true._ Because you do not know which of these is correct, you should proceed by cases; that is, you should do two proofs. In the first case (proof) you assume that _D_ is _true_ and prove that _B is true_; in the second case you assume that _E is true_ and prove that _B is true._

12.2   If the contradiction method is used on each of the following problems, what technique will you use to work forward from NOT B?

a.   A implies (C AND D).

Remember that in the contradiction method, you prove that "_A implies B_" by assuming _A and NOT(B)_, and based on it you prove that _NOT(A)_ holds. Thus, there is a contradiction and the assumption NOT(B) was wrong – and you can conclude that "_A implies B_" is true.

In our case, "_B_" is "_C AND D_". Therefore, our assumptions are:

$$NOT\ ``B" = NOT(C\ AND\ D) = (NOT\ C\ OR\ NOT\ D).$$

So in the initial hypothesis, we have:

$$NOT\ C\ OR\ NOT\ D.$$

And so if we work by the forward method, we will have to apply the _proof by cases_ approach.

163

Let us assume that *NOT C* is true, and then see what happens (that is, try to prove that *NOT A* is also true).

Let us assume that *NOT D* is true, and then see what happens (that is, try to prove that *NOT A* is also true).

Once we have succeeded in the proof starting from *NOT C* or *NOT D*, we are done, as explained in the *proof by cases* approach.

b.   A implies [(Not C) And (Not D)]

Remember that in the contradiction method, you prove that "*A implies B*" by assuming *A and NOT(B)*, and based on it you prove that *NOT(A)* holds. Thus, there is a contradiction, and the assumption NOT(B) was wrong . Therefore you can conclude that "*A implies B*" is true.

In our case, "B" is "(Not C) And (Not D)". Therefore, our assumptions are:

$$NOT\ "B" = NOT(NOT\ C\ AND\ NOT\ D) = NOT(NOT\ C)\ OR$$
$$NOT(NOT\ D) = C\ OR\ D.$$

So, in the initial hypothesis we have:

$$C\ OR\ D.$$

If we work by the forward method, we will have to use the proof by cases approach.

Let us assume that *C* is true, and see what happens (that is, try to prove that *NOT A* is also true).

Let us assume that *D* is true, and see what happens (that is, try to prove that *NOT A* is also true).

Once we have succeeded in the proof, starting from *C* or *D*, we are done, as explained in the *proof by cases* approach.

12.4   Explain where, why and how the proof by cases approach is used in the condensed Proof of the Proposition below.

Proof of Proposition: *To show that S = T, it is shown that S is a subset of T, and T is a subset of S.*

*To see that S is a subset of T, let x element of S. (Remember that he use of the word **"let"** in the condensed proofs frequently indicates that the choose method is being invoked.)*

Consequently, $x^2 - 3x + 2 \leq 0$ and so $(x-2)(x-1) \leq 0$. This means that either $x-2 \geq 0$ and $x-1 \leq 0$ or else $x-2 \leq 0$ and $x-1 \geq 0$. The former cannot happen because, if it did, $x \geq 2$ and $x \leq 1$. Hence it must be that $x \leq 2$ and $x \geq 1$, which means that x element of T.

Remember that the "proof by cases" method states that to prove that *(C or D) implies B*, we proceed by the cases approach.

Firs,t we divide the proof in two steps, and try to prove that:

C implies B
Or
D implies B.

If you are able to prove just one of these two cases, then you have proved that *(C or D) implies B*, as we wanted.

Proof of Proposition: To show that S = T, it is shown that S is a subset of T and T is a subset of S.

Note that two sets are equal *if and only if* each set is included in the other.

To see that S is a subset of T, let x element of S.

We will try to prove that *S* is a subset of *T*, and that *T* is a subset of *S*.

Let us consider *x* to be an element of *S*, and will try to prove that *x* is also in set *T*.

(The use of the word "let" in the condensed proofs frequently indicates that the choose method is being invoked.)

Yes! The choose method was used when we selected an x element of S, with the intention of proving that x is also an element of T.

(proof) Consequently, $x^2 - 3x + 2 \leq 0$ and so $(x-2)(x-1) \leq 0$.

Let us assume that S is defined as being the set of real such that $x^2 - 3x + 2 \leq 0$.

Thus, *x is in S* means that $x^2 - 3x + 2 \leq 0$. That is, that

$$(x-2)(x-1) \leq 0.$$

This means that either x-2 $\geq$ 0 and x-1 $\leq$ 0, or x-2 $\leq$ 0, and x-1 $\geq$ 0.

Observe that (x-2)(x-1) $\leq$ 0 may happen if one of the following happens:

(x-2) and (x-1) have opposite signs, that is, (x-1) $\leq$ 0 and (x-2) $\geq$ 0 OR (x-1) $\geq$ 0 and (x-2) $\leq$ 0.

Now we want to prove the following hypothesis:

(x-1) $\leq$ 0 and (x-2)$\geq$0 OR (x-1) $\geq$ 0 and (x-2) $\leq$ 0 implies the conclusion: x is in T.

Here is where we invoke the proof by cases method. To prove that *(C or D) implies B*, we will try to prove that *C implies B*, or that *D implies B*. The former cannot happen, because if it did, $x \geq 2$ and $x \leq 1$.

So first we try to prove that *(x-2) $\geq$ 0 and (x-1) $\geq$ 0* implies x is in T. This is not possible, as *(x-2) < (x-1)*, so it cannot be *(x-2) $\geq$ 0 $\geq$ (x-1)*.

Hence it must be that $x \leq 2$ and $x \geq 1$, which means that x is an element of T.

12.6

Proof by Elimination: Suppose the key words "either/or" arise in the conclusion of a proposition whose form is "*A implies C OR D*". With the forward-backward method, you assume *A is true* and need to conclude that either *C is true* or else *D is true*. The only question is whether you should try to show that *C is true* or whether you should try to show that *D is true*.

Explain where, why, and how a proof by elimination is used in the following proof.

Proposition: *If* $i = (-1)^{1/2}$ *and* $a + bi$ *and* $c + di$ *are complex numbers of which* $(a + bi)(c + di) = 1$. *them* $a \neq 0$ *or* $b \neq 0$.

Proof: *Because* $(a+bi)(c+di) = 1$, $ac + adi + bci - bd = 1$.
That is, $ac - bd = 1$ *and* $ad + bc = 0$. *If* $a = 0$, *then* $-bd = 1$.
It then follows that $b \neq 0$ *and so the proof is complete.*

Note that in the elimination method we want to prove that "*A implies C OR D*" by selecting C or D and proving "*A implies C*" or "*A implies D*" (because by proving just one of the last statements, we would be proving the first one).

    Because $(a+bi)(c+di) = 1$, $ac + adi + bci - bd = 1$.

We know that $(a+bi)(c+di) = 1$.

We also know that $(a+bi)(c+di) = ac + adi + dci + bd*i^2$.

We know that the complex number, $i^2 = -1$, and so $bd * i^2 = -bd$.

Therefore, $(a+bi)(c+di) = ac + adi + dci - bd$.

And as we already know that $(a+bi)(c+di) = 1$, we have $ac + adi + dci - bd = 1$. Then, $ac - bd = 1$ and $ad + bc = 0$.

This is a very simple property of complex numbers. The property is that if you have the complex numbers $A + Bi = C + Di$ (where A, B, C, D are real numbers and "i" is a complex), then $A + Bi = C + Di$ implies that $A = C$ and $B = D$.

In our case, the equality, $ac + adi + dci - bd = 1$, can be rewritten as $(ac - bd) + (ad + dc)i = 1 + 0 * i$.

Which means that $(ac - bd) = 1$ and $(ad + dc) = 0$, and if $a = 0$, then $-bd = 1$.

Now, we will begin using the elimination method. We want to prove that $a \neq 0$ or $b \neq 0$. So it is sufficient to prove just one of the two assertions to complete the proof by elimination method.

Now, suppose that $a \neq 0$ holds. Then, we have completed the proof. So the remaining case to be considered is the case $a = 0$, and that implies $-bd = 1$ (because $ac - bd = 1$)

It then follows that $b \neq 0$ and so the proof is complete. And as $-bd = 1$ implies $b \neq 0$, then if $a = 0$, then $b \neq 0$ proving that at least one of a or b must be different to 0, as we wanted.

12.10 Provide an analysis of proof for the following condensed proof.

Proposition: *If n is a positive integer, then either n is prime, or n is a square, or n divides (n-1)*

Proof: *If $n = 1$ then $n = 1^2$ is a square and the proposition is true. Similarly, if $n = 2$, then n is prime and again the proposition is true. So suppose that $n > 2$ is neither prime nor a square. Because $n > 2$ is not prime, there are integers a and b with $1 < a < n$ and $1 < b < n$ such that $n = ab$. Also, because n is not square, $a \neq b$. This means that a and b are integers with $2 \leq a \neq b$. That is, a and b are two different terms of $(n-1)(n-2)....1$. Thus, $ab = n$ divides $(n-1)$!*

We will use the elimination method to complete the proof, because the proof is in the form, *A implies (B or C or D)*. Therefore, it is sufficient to prove, "*A implies B*", "*A implies C*" or "*A implies D*" to complete the proof.

(proof) If n = 1 then n=1² is a square and the proposition is true.

In the case $n = 1$, the statement is trivial because $n = 1^2$ is a square. Therefore, the fact that n is positive, implies n is a square for the case n = 1.

The proposition holds for $n = 1$ due to the elimination method, because it is sufficient to prove that *"n is a positive integer, implies n is a square"*, to prove the complete proposition.

(proof) Similarly, if n=2, then n is prime and again the proposition is true.

In the case $n = 2$, the statement is obvious again because $n = 2$, and 2 is a prime number. Therefore, n is positive implies n is prime for the case n = 2.

The proposition holds for $n = 2$ due to elimination method, because it is sufficient to prove that *"n is a positive integer, implies n is prime"*, to prove the complete proposition.

(proof) So suppose that n >2 is neither prime nor a square.

Let us consider the case $n > 2$. We want to prove that it is a square or it is a prime number or it divides (n-1)!

Again, by the elimination method, it is sufficient to prove that one of the three conclusions holds to prove that the conclusion is correct. Therefore, if it is a square, then we are done, and have proved that n verifies the proposition. If it is a prime, then we are done, and have proved that n verifies the proposition. The remaining case to consider is where all *n's that are not square or prime numbers.* We should be able to prove that *n divides (n-1),* because by the elimination method, we need at least one of the these three: *"n is prime", "n is a square", "n divides (n-1)!"* to be verified.

Let us suppose that n is not a square, nor a prime number, and let us try to prove that *n* is such that *n divides (n-1)!*

(proof) Because n > 2 is not prime, there are integers a and b with 1 < a < n and 1 < b < n such that n = ab.

This is the definition of being a prime number. By the definition of a prime number, $n > 2$ is a prime number if and only if $n = ab$, where a and b are two integers > 1 and $< n$.

(proof) Also, because n is not square, a ≠ b.

We also know that a and b are different, because n is not a square.

(proof) This means that a and b are integers with 2 ≤ a ≠ b. That is, a and b are two different terms of (n-1) (n-2)....1.

True! Since both are $< n$ and > 1. Therefore, they must be terms of (n-1)! And as they are different (not repeated), they are a divisor of (n-1)!

(proof) Thus, ab=n divides (n-1)!

Therefore, the conclusion is that, ab / (n-1)!, and n / (n-1)!, as we needed to prove.

12.12   Prove that if S and T are subsets of a universal set U, then (S Intersection T) ^{c} = S ^{c} Union T ^{c}, where X^c = { x element of U : x not element of X}. Indicate clearly where the either/or method arise.

We have to prove that if S and T subsets of U, then *(S intersection T)^c = S ^{c} U T ^{c}* (c means "complement").

Now, as $A = B$ can be expressed as A included in B, and B included in A, we have to prove the following:

IF S and T are subsets of U, universal set
THEN
(S intersection T)' included in (S ^{c} U T ^{c})
AND
(S ^{c} U T ^{c}) included in (S intersection T) ^{c}

Therefore, we must prove two things:

$$(S \text{ intersection } T)^c \text{ included in } (S^c \cup T^c)$$
$$AND$$
$$(S^c \cup T^c) \text{ included in } (S \text{ intersection } T)^c$$

Let us try to prove that *(S intersection T)* ^{c} included in *(S ^{c} U T ^{c})*. We want to prove the following:

if x is in (S intersection T) ^{c}, *then x is in S ^{c} U T ^{c}*

That is, if x is not in the *(S intersection T)*, then x is in $S^c \cup T^c$. (remember that being in A ^{c} means being not in A).

If x is not in the *(S intersection T)*, then x is not in S, or x is not in T. (this is the meaning of being in $S^c \cup T^c$).

Now applying the elimination method, it is sufficient to prove that one of the following is true:

1) if x is not in (S intersection T), then x is not in S.
OR
2) if x is not in (S intersection T), then x is not in T.

Now, if x is not in S, we would have completed the proof, since (1) would be verified.

Let us suppose that x is in S. Then x cannot be in T, because if it was, it would be in the *(S intersection T)*, which is false by hypothesis. So (2) is verified.

So we have proved that, in all cases, at least (1) or (2) is verified, which means that the first assertion has been proved:
(S intersection T) ^{c} is included in *(S ^{c} U T ^{c})*.

Let us now prove, that *(S ^{c} U T ^{c})* is included in *(S intersection T)* ^{c}.

If x is in the *(S' U T ^{c})*, then x is *in (S intersection T)* ^{c}. That is, if x is not in S or x is not in T, then x is not in *(S intersection T)*.

But this is obviously true, because if x is not in S, then it cannot be in the *(S intersection T)*. And if it is not present in T, then neither can it be in *(S intersection T)*.

In our conclusions above we used the *Proof By Cases method* in the forward process to conclude that if x is in $(S^c \cup T^c)$ then x is in the *(S intersection T)* ^{c}, and we needed to prove *(S ^{c} U T ^{c}) is included in the (S intersection T)* ^{c}

12.14 Convert each of the following statements into an equivalent statement having a quantifier.

MAX/MIN Methods: this technique is used in problems dealing with maxima and minima. Suppose that S is a nonempty set of real numbers, having both a largest and smallest member. For a given real number x, you might be interested in the position of the set S relative to the number x.

For example, you might want to prove one of the following statements:

1. All of S is to the right of x
2. Some of S is to the left of x
3. All of S is to the left of x
4. Some of S is to right of x

In mathematical problems these four statements are likely to appear as (respectively):

(a) min {s: s Element of S} $\geq$ x.
(b) min {s: s Element of S} $\leq$ x.

a.   The maximum of a function $f(x) \leq y$, where x is a real number with $0 \leq x \leq 1$ and y is a given real number.

We want the maximum value of a function $f(x) \leq y$, where x is a real number between 0 and 1 and y is given.

So we want that, *for all x such that $0 \leq x \leq 1$, $f(x) \leq y$.*

172

b. The minimum of a function f(x) ≤ y, where x is a real number with $0 \le x \le 1$ and y is a given real number.

This case is trickier. We will look at it more carefully. We want the minimum value of $f(x) \le y$, where x is between 0 and 1.

We want the *f(min value)* $\le y$, but there is nothing stated about the other values of x. It may happen that $f(x) \ge y$ for some values of x.

Therefore, we want for some values of x, $f(x) \le y$, or in a more formal notation, "*there is an x such that $0 \le x \le 1$ that verifies f(x) ≤ y.*"

174

Chapter 13
Summary

13.2   For each of the following state how the technique you chose as the way to complete the proof would be applied to that problem. That is, indicate what you would assume, what you conclude, and how you would go about doing it.

a.   If p and q are odd integers, then the equation $x^2 + 2px + 2q = 0$ has no rational solution for x.

We want to prove that $x^2 + 2px + 2q = 0$ has no rational solutions for x (given that p and q are odd integers). To do the proof, we could use the contradiction method: that is, we could assume that there is a rational solution for $x^2 + 2px + 2q = 0$, and conclude from this assumption that p or q must be an even integer, or not an integer. But we would find a contradiction proving that there are no rational solutions.

Let us assume that $x^2 + 2px + 2q = 0$ has a rational solution. Then there is a rational number m/n such that

$$(m/n)^2 + 2p * (m/n) + 2q = 0.$$

We can assume that m/n is not reducible, that is, there are no common divisors for m and n. Now if we multiply both sides of the equation by n^2, we get the following:

$$m^2 + 2pmn + 2qn^2 = 0$$
$$m^2 = 2pmn + 2qn^2.$$

And so m is an even number.

Since m is even, m^2 is a multiple of 4, and $2pmn$ is also a multiple of 4, and $2qn^2$ must also be a multiple of 4. However, since q is odd, then n^2 *is* even, and n also is even.

We have found that m is even, and n is even, which is in contradiction with how they were selected, as non-reducible. Therefore, we can conclude that our original assumption was wrong, and there is no rational solution for x. Hence, we have proven the desired assertion by the contradiction method.

b.   For every integer n ≥ 4, n! > n²

We will try to prove it by the induction method.

1) **BASE STEP**:

To prove for the case for $n = 4$, that is, we have to prove that $4! > 4^2$, which is true because $4! = 4*3*2*1 = 24$, and $4^2 = 16$.

2) **INDUCTION STEP**:

If $n! > n^2$ then $(n+1)! > (n+1)^2$ *(for n ≥ 4).*

Let us try to prove it, and let us assume that $n! > n^2$. We will start by going forward trying to prove the desired conclusion.

If we multiply both sides by *(n+1)*, and we get the following:

$$(n+1) * n! > (n+1)*n^2$$

$$(n+1)! > (n+1)*n^2$$

Since $n^2 = n*n > 4$ if $n ≥ 4$, then we have:

$$n^2 > 4n = n+n+n+n > n+1$$

Therefore,
$$n^2 > n+1$$

And so,
$$(n+1) * n^2 > (n+1)^2$$

Hence, we have proved that $(n+1)! > (n+1)^2$.

c.   If f and g are convex functions, then f + g is a convex function.

We can use the definition of being *convex*, and we can use the forward method to prove the problem based on the hypothesis.

We can start from the statement that *f and g* are convex. Then applying the definition, we get the following:

$$\text{for all } x, y \text{ and for all } t \ (0 \le t \le 1)$$

$$f(tx + (1-t)y) \le t * f(x) + (1-t) * f(y).$$

And

$$g(tx + (1-t)y) \le t * f(x) + (1-t) * f(y).$$

Let us try to prove the desired conclusion based on the following:

By definition of $f+g$

$$(f + g)(tx + (1-t)y) = f(tx + (1-t)y) + g(tx + (1-t)y)$$

$$\le tf(x) + (1-t)f(y) + tg(x) + (1-t)g(y)$$

$$= t(f(x) + g(x)) + (1-t)(f(y) + g(y))$$

$$= t(f+g)(x) + (1-t)(f+g)(y)$$

And so we have proved that

$$(f+g)(tx + (1-t)y) \le t(f+g)(x) + (1-t)(f+g)(y).$$

And we have proved that $f + g$ is a convex function.

d. If a,b and c are real variables, then find the maximum value of ab + bc + ac subject to the condition that $a^2 + b^2 + c^2 = 1$ is less than or equal to 1.

We will try to prove this problem by using the contradiction method, ie, assume that the conclusion is false, and then try to find a contradiction with the hypothesis.

The *hypothesis* is that $a^2 + b^2 + c^2 = 1$.

The *conclusion* is that the maximum value of $ab + bc + ac$ *is* ≤ 1.

Let us assume that the conclusion is false, that is, that the maximum value of $ab + bc + ac$ *is* > 1.

Now consider the following:

$a^2 + b^2 + c^2 - ab - bc - ac =$

$$= (a^2 + b^2)/2 - ab + (b^2 + c^2)/2 - bc + (a^2 + c^2)/2 - ac$$

$$= (a^2 - 2ab + b^2)/2 + (b^2 - 2bc + c^2)/2 + (a^2 - 2ac + c^2)/2$$

$$= (a - b)^2 / 2 + (b - c)^2 / 2 + (a - c)^2 / 2$$

$$\geq 0 \text{ because it is the sum of three squares}$$

So we have proved that

$$a^2 + b^2 + c^2 - ab - bc - ac \geq 0$$

Implying that

$$a^2 + b^2 + c^2 \geq ab + bc + ac$$

$$ab + bc + ac > 1$$

$$a^2 + b^2 + c^2 > 1$$

But this is in contradiction with the hypothesis $a^2 + b^2 + c^2 = 1$. Therefore, it means that our initial assumption that the conclusion was false is wrong. The conclusion is true, and so we have proved that the assertion by the contradiction method.

e. In a plane, there is one and only one line perpendicular to given line l through a point P on the line.

Here we can use the *indirect uniqueness* method, that is, we can assume that there are two perpendicular lines to line l by the point P (assume p_1 and p_2 are perpendicular to line l), and based on it we can prove that p_1 and p_2 must be equal.

For example, we can use the fact that if p_1 and p_2 are perpendicular to l in the plane, then p_1 // p_2. Since we know that p_1 and p_2 are parallel, and have a common point P, then they must be coincident, and so $p_1 = p_2$, as we wanted to prove.

f. If f and g are functions such that (1) for all real numbers x, $f(x) \leq g(x)$ and (2) there is no real number M such that for all x, $f(x) \leq M$, then there is no real number $M > 0$ such that for all real numbers x, $g(x) \leq M$.

We can try to prove this problem by the contradiction method, that is, we can assume that the conclusion is false, and then try to prove that this is in contradiction with the hypothesis.

The hypothesis is that:

for all real numbers x, (1) $f(x) \leq g(x)$ and (2) there is no real number M such that for all x, $f(x) \leq M$.

The conclusion is that:

There is no real number $M > 0$ such that for all real numbers x, $g(x) \leq M$.

We start by assuming, using the contradiction method, that the conclusion is false, that is

THERE IS a real number $M > 0$ such that for all real numbers x, $g(x) \leq M$.

We know by hypothesis (1) that $f(x) \le g(x)$ for all x, so we have that, for all real numbers x, $f(x) \le g(x) \le M$, and so we have proved the following:

THERE IS a real number M > 0 such that for all real numbers
$$x, f(x) \le M.$$

But this is in contradiction with hypothesis 2. Therefore, we are done. We have found a contradiction with the hypothesis, which means that we were wrong in contradicting the conclusion, and that the original statement was true.

g.   If f and g are continuous functions at the point x, then so is the function f+g.

We can try using the forward method to prove it. We can start from the hypothesis that

f and g are continuous functions

And by applying the definition of being *continuous*, and then we have the following:

For all x, $\lim_{(h \to)} f(x+h) = f(x)$ and $\lim_{h \to 0} g(x+h) = g(x)$.

Now adding the limits

$$\lim_{h \to 0} (f(x+h) + g(x + h)) = f(x) + g(x).$$

And using the definition of addition of functions, that is

$$(f+g)(x) = f(x) + g(x)$$

$$\lim_{h \to 0} (f+g)(x + h) = (f+g)(x).$$

This is the definition of a function $f+g$ being continuous at x. Therefore, we have proved that

if f is continuous at x and g is continuous at x, then f+g is continuous at x

We have proved that "*if f and g are continuous, then f+g are continuous.*"

h.   If f and g are continuous functions at the point x, then for every real number e>0, there is a real number gamma>o such that for all real numbers y with |x-y| < gamma, |f(x) + g(x) - (f(y) + g(y) | < e.

We can use the contradiction method to prove this problem, that is, we can assume that the conclusion is false to reach a contradiction with the hypothesis.

The hypothesis is that:

"f and g are continuous functions at the point x".

The conclusion is that:

"for every real number e > 0 there is a real number gamma > 0 such that for all real numbers y with $|x - y| <$ gamma, $|f(x) + g(x) - (f(y) + g(y))| < e$".

We can assume that the conclusion is false. Therefore,

NOT(for every real number e > 0 there is a real number gamma > 0 such that for all real numbers y with $|x - y| <$ gamma, $|f(x) + g(x) - (f(y) + g(y))| < e$).

That is,

THERE IS a real number e > 0 such that FOR ALL real number gamma > o THERE IS y with $|x - y| <$ gamma such that $|f(x) + g(x) - (f(y) + g(y)) | \ge e$.

Now let us consider that e > 0 verifies the previous assertion. Therefore, for this e > 0, we have that the following is true:

FOR ALL real number gamma > o THERE IS y with $|x - y| <$ gamma such that $|f(x) + g(x) - (f(y) + g(y))| \ge e$.

That is,

FOR ALL real number gamma > 0 THERE IS y with $|x - y| <$ gamma such that $|(f+g)(x) - (f+g)(y)| \ge e$.

And for *gamma* = e, we have the following:

THERE IS y with $|x - y| < e$ such that $|(f+g)(x) - (f+g)(y)| \geq e$.

Which means that *(f+g)* is NOT a continuous function.

However, we have proved previously in part (g), that if *f and g* are continuous, then *(f+g)* is continuous. Therefore, if *(f+g)* is not continuous, this implies that at least *f or g* is NOT a continuous function, which is in contradiction with our initial hypothesis.

Therefore, we have reached a contradiction, which means that we cannot assume that the conclusion is false. We can conclude that the original statement is correct, and we have completed the proof by the contradiction method.

i. If f is the function defined by $f(x) = 2^{-x} + (x/2)$ then there is a real number x* between 0 and 1 such that for all y, $f(x^*) \leq f(y)$.

We can proceed by the construction method. We will try to find (build) an x* between 0 and 1 such that $f(x^*) \leq f(y)$ for all y.

We are trying to find an *x* = MINIMUM* of the function *f* in the interval *(0, 1)*. To do this, we have to compute the derivative of *f*.

$$f'(x) = -ln(2) * 2^{-x} + \tfrac{1}{2}.$$

We are searching for the value x^* such that

$$f'(x^*) = 0$$

Therefore, it must be that

$$-ln(2) * 2^{-x^*} + \tfrac{1}{2} = 0$$

That is,

$$ln(2) * 2^{-x^*} = 1/2$$
$$2^{-x^*} = 1/(2\,ln(2))$$
$$-x^* = ln(1/2\,ln(2))$$
$$x^* = -ln(1/2\,Ln(2)) = ln(2\,Ln(2)).$$

So, we have found an x^* such that for all y in $(0, 1)$, $f(x^*) \leq f(y)$.

13.4 Suppose the forward-backward method is used to start each of the following proofs. List all of the techniques that are likely to be used subsequently in the proof.

a. (C AND D) implies (E or F)

We could use the elimination method in the backward process (applied to *E* or *F*) to prove that *"NOT C or NOT D"*.

We could also use the contradiction method, starting from *NOT(E or F)* and try to prove *NOT(C and D)*, that is, starting from *NOT E* and *NOT F* and try to prove *NOT C or NOT D*.

b. (C or D) implies (E and F)

Here we should use the proof by cases applied to *(C or D)* in the forward process starting at *"NOT E or NOT F"*.

We could also use the contradiction method, starting from *NOT (E and F)* and try to *prove NOT(C or D)*, that is, starting from *(NOT E or NOT F)* and try to *prove (NOT C and NOT D)*.

c. If X is an object such that for all objects Y with a certain property, something happens, then there is an object Z with a certain property such that something else happens.

We have to prove that *"there is"* such an object Z. Therefore, we should construct this object using the construction method.

d. If for all objects X with a certain property, something happens, then there is an object Y with a certain property such that, for all objects Z with a certain property, something else happens.

We want to prove a *"there is"* such an object Y. Therefore, we should construct it using the construction method. We construct an object *Y* such that *"for all objects Z with a certain property, something else happens"*. Therefore, we select a given *Z* with a certain property, and try to prove that for this *Z*, *something else happens*.

13.5   Repeat problem 13.4 assuming that the contrapostive method is used to start each proof.

a.   (C AND D) implies (E or F)

With the contrapositive method, we want to prove the following:

>*NOT(E or F) implies NOT (C AND D)*

That is,

>*(NOT E and NOT F) implies (NOT C or NOT D)*

Therefore, we should use the *elimination method* in the backward process and apply it to *"NOT C or NOT D"*.

b.   (C or D) implies (E and F)

With the contrapositive method, we want to prove the following:

>*NOT (E and F) implies NOT(C or D)*

That is,

>*(NOT E or NOT F) implies (NOT C and NOT D)*

Here we should use the *proof by cases* and apply it
"NOT E or NOT F" in the forward process.

c.   If X is an object such that for all objects Y with a certain property, something happens, then there is an object Z with a certain property such that something else happens.

The contrapositive of the following:

>*NOT (there is an object Z with a certain property such that*
>*something else happens)*

Implies that

>*NOT(X is an object such that for all objects Y with a certain*
>*property, something happens).*

184

That is,

FOR ALL object Z with a certain property, NOT(something else happens).

Implies that

X is an object such that THERE IS   Y with a certain property such that NOT (something happens).

We have to prove that *"there is"* such an object Y. Therefore, we should construct this object Y, given the X, using the construction method.

<u>d.</u>　If for all objects X with a certain property, something happens, then there is an object Y with a certain property such that, for all objects Z with a certain property, something else happens.

With the contrapositive method, we want to prove the following:

NOT(there is an object Y with a certain property such that, for all objects Z with a certain property, something happens)

Which implies that

NOT(for all objects X with a certain property, something happens)

That is,

FOR ALL objects Y with a certain property, THERE IS an object Z with a certain property such that NOT(something happens)

Which implies that

THERE IS an object X with a certain property such that NOT(something happens)

We want to prove that *"there is"* such an object X. Therefore, we should construct an object X such that, *the something does not happen.*

185

Index